NATURE

Some of the most exciting, alarming and dramatic developments of our time involve nature (human and non-human). These range from xeno-transplantation to climate change to the Human Genome Project. Geographers have long counted nature among their principal objects of analysis. Indeed, it's fair to say that the nature of geography as a subject is intimately linked to the nature that geographers study.

This book offers an incisive introduction to the nature that geographers study and, as a corollary, the nature of geography as a discipline. It is written for researchers, degree students and their teachers. It is the first book to bring the diverse aspects of the nature that geographers examine – and the myriad ways they study it – within a common frame of analysis. The book treats geography as an active producer of societal understandings of nature. Since their veracity can never be established in any absolute way, *Nature* treats these knowledges as ideas about nature that must battle it out to win the minds and hearts of students, funding bodies, governments and all those other organisations and constituencies who are interested in the knowledge that geographers produce. These ideas, in one sense, create the 'realities' they purport to describe and explain. The knowledges of nature that geographers produce must, therefore, be seen as part of a high-stakes contest over how we understand and act towards those myriad things we label 'natural'. This contest has implications for us all, as well as for the non-human world.

Nature is an advanced introduction to its topic. For students, it aims to inform and to challenge by showing that nature is not what it seems to be. For geography teachers and researchers, *Nature* brings together ideas and arguments hitherto compartmentalised into geography's three main parts (human, physical and environmental geography). In so doing it offers fresh insights into one of the discipline's most familiar, yet elusive, objects of analysis, policy formation and moral concern.

Noel Castree is a Professor in the School of Environment and Development at Manchester University. He is the co-editor of *Remaking Reality* (1998) and *Social Nature* (2001).

D0144837

Key Ideas in Geography

Series editors: Sarah Holloway, Loughborough University and Gill Valentine, Leeds University

The Key Ideas in Geography series will provide strong, original, and accessible texts on important spatial concepts for academics and students working in the fields of geography, sociology and anthroplogy, as well as the inter-disciplinary fields of urban and rural studies, development and cultural studies. Each text will locate a key idea within its traditions of thought, provide grounds for understanding its various usages and meanings, and offer critical discussion of the contribution of relevant authors and thinkers.

Published

Nature
NOEL CASTREE

Forthcoming

Home
ALISON BLUNT AND
ROBYN DOWLING

Scale
ANDREW HEROD

City
PHIL HUBBARD

Landscape
JOHN WYLIE

Environment
GEORGINA ENDFIELD

NATURE

Noel Castree

Routledge
Taylor & Francis Group

LONDON AND NEW YORK

First published 2005
by Routledge
2 Park Square, Milton Park, Abingdon, Oxon OX14 4RN

Simultaneously published in the USA and Canada
by Routledge
270 Madison Ave, New York NY 10016

Routledge is an imprint of the Taylor & Francis Group

© 2005 Noel Castree

Typeset in Joanna and Scala Sans by
Keystroke, Jacaranda Lodge, Wolverhampton
Printed and bound in Great Britain by
TJ International Ltd, Padstow, Cornwall

British Library Cataloguing in Publication Data
A catalogue record for this book is available from the British Library

Library of Congress Cataloging in Publication Data
A catalog record has been requested

ISBN 0–415–33904–9 (hbk)
ISBN 0–415–33905–7 (pbk)

'If I ask about the world, you can offer to tell me how it is under one or more frames of reference; but if I insist that you tell me how it is apart from all frames, what can you say?'

(Nelson Goodman 1978: 13)

Contents

Illustrations and tables

MAPS

TABLES

BOXES

ACKNOWLEDGEMENTS

Writing this book has, in equal measure, been difficult, surprising and rewarding. Of all the key concepts examined in the series to which this book belongs, nature is arguably the most difficult to understand. Promiscuous in its meanings and referents, I quickly realised that to discuss the 'nature' that geographers have studied necessarily entails discussing 'the nature of geography' as a discipline (rather than just a few of its constituent parts). To perform this daunting double-task in a clear, succinct way I needed to find a parsimonious means of presenting ideas, insights and arguments from across geography's many research communities. In an advanced introduction of this kind, it was important to ensure fair coverage of nature's multifarious manifestions in geographical discourse while avoiding a superficial gloss of the issues that such a breadth of coverage threatens to produce. It took a very long time – and many headaches – before I settled on the six-chapter structure found here. Given the enormity and complexity of my topic, I'm hopeful that this structure organises a potentially unruly mass of information into some sort of coherent analysis.

I need to thank Sarah Holloway and Gill Valentine for asking me to write this book. When they first approached me in mid-2003 I felt there was little new to be said about nature. After all, the concept had been famously analysed by Raymond Williams and, more recently, by Kate Soper. What's more, the week after I signed the contract for this book, I encountered John Habgood's *The Concept of Nature*. Unbeknownst to me, Habgood's insightful little text had been published late in 2002. It was the latest in a long line of books about nature, whose distinguished authors included R.G. Collingwood, C.S. Lewis and Alfred North Whitehead. This confirmed my initial suspicion that I was wasting my time repeating what others had done far better than I could ever do. But then I recalled that I was being asked to write not about nature in general but about how it's been understood in the discipline of *geography* – a discipline that, in its own specific ways, has influenced (as well as been influenced by) wider understandings of nature since its formal constitution as a research and teaching subject in the late nineteenth century.

This changed everything. While the writings of Williams, Soper, Habgood and others dissect the concept of nature with forensic precision, what they fail to do is anchor the concept in the various sites, situations and institutions where its meaning and referents are 'fixed' or contested. Ideas about nature (and, indeed, about everything else in our world) do not exist on the head of pin nor are they abstract entities that somehow 'touch down' uniformly across time and space. Rather, they are produced by myriad knowledge-communities who possess sometimes similar (and sometimes different) outlooks on nature. What we call 'societal' understandings of nature are, in reality, 'local' understandings that have leaked out from the sites of their production so that numerous people come to accept them as valid. These sites include universities. Professional geographers are one of the many knowledge-communities that have spent (and still spend) a great deal of time producing and disseminating knowledge about nature. Yet no one has attended to the specific ways that geographers as a whole understand nature. Instead, one typically encounters texts that offer only partial coverage – like my own co-edited *Social Nature*, a book that examines how nature is understood by critical human geographers (Castree and Braun 2001).

A book can only be written with the assistance of many friends, colleagues and interlocutors. *Nature* could not have been conceived or completed without the help of Bruce Braun, David Demeritt and Ali Rogers. In different ways each of them have influenced the substance of this book and I continue to value their friendship greatly. Dave and Ali deserve special thanks for volunteering to read the manuscript. Their detailed and wise comments proved invaluable and *Nature* is a better book for them. Ali also suggested the title for Chapter 1, while Sarah Holloway's careful editing saved me from some egregious errors. I must also thank, without implicating, the two anonymous reviewers of my original book proposal. Their suggestions have certainly helped to improve the content and structure. Likewise, Vinny Pattison (now a PhD student) made very constructive comments on the first draft of the book. Here at Manchester, Clive Agnew, Mike Bradford,

Neil Coe, Peter Dicken, Chris Haylett, John Moore, Brian Robson, Fiona Smyth, Kevin Ward and Jamie Woodward have all, in different ways, made working in the School of Geography (now School of Environment and Development) a pleasure. The School's Research and Graduate Committee granted me sabbatical leave in 2004. This speeded the completion of *Nature* and I'm grateful for that. I'm also indebted to students of GE3070 who, even though they didn't realise it, had to endure a course that I would now teach quite differently having written this book.

Sometimes one only appreciates the full significance of a relationship years after the event. Though our paths cross only infrequently these days, the enduring influence of my PhD thesis advisers – Derek Gregory and Trevor Barnes – is evident in these pages. *Nature* tries to dissect its topic clinically yet without being dry or forbidding. If it succeeds in this then it's due in no small measure to Trevor and Derek's influence. As ever, I wish to thank Nick Scarle for his exemplary skills with the book's figures, maps and tables. Finally, the editorial and production teams at Routledge deserve a big thank-you: they've given me all the right help at all the right times over the last year.

Despite my reluctance to talk about nature in any simple or straight-forward way, I must nonetheless acknowledge that natural miracles do occur. In late February 2004 I became a father. Even if I understood all the biological intricacies that made this possible, I shall never get over my astonishment, gratitude and wonder that such a thing can happen at all. I dedicate this book to Marie-Noel and Thomas. They are, naturally, the loves of my life.

The author and publisher would like to thank the following for granting permission to reproduce material in this book: Blackwell Publishers for Figures 1.3 and 4.6, both reproduced from the *Transactions of the Institute of British Geographers*, and Figure 3.2 from N. Castree and B. Braun (2001) *Social Nature*; Jack Kloppenburg for Figure 3.1, reproduced from *First the Seed* (Cambridge University Press, 1988); Cambridge University Press for Figures 4.3 and 4.4, reproduced from S. Schumm (1991) *To Interpret the Earth* and von Engelhardt et al. (1988) *Theory of Earth Science* respectively; *American Journal of Science* for Table 4.1; and Harvard University Press for Figure 4.1, reproduced from B. Latour (1993) *We Have Never Been Modern*.

Preface

Read this

Nature is one of the most widely talked about and investigated things there is. Today, one routinely hears apocalyptic declarations about the 'end of nature', plangent injunctions to 'save' nature or else brave-new-world claims that nature can be improved upon and its 'imperfections' eliminated. Such rhetoric implies that nature is currently 'on the agenda' as never before. But this implication is false. Nature has *always* been a major issue for societies worldwide. What's changed is the way we talk about and act towards those things conventionally called 'natural'. It is a truism that 'facts never speak for themselves'. Nature is a living testament to this. Historically, ideas about nature have changed dramatically. Yesterday's 'truths' about nature often seem absurd to us in the here and now. Nature continues to be understood in a multitude of ways, many of them incompatible. Indeed, the struggle to get a 'proper' understanding of nature is one of the defining struggles of any era.

Geography is one of several disciplines that has played a role in influencing societal understandings of nature. Geographical teaching and research have shaped the nature-imaginaries of countless people since geography came into existence as a university subject in the late nineteenth century. It is one of several disciplines – the others being the physical, medical and engineering sciences, for the most part – that are responsible for producing 'expert knowledges' about natural phenomena. By the same token, academic disciplines are just one of several domains in which knowledges of nature are produced. Other domains include the mass media (television, newspapers and magazines), the entertainment industry (think of *Jurassic Park* or *The Hulk*, two morality tales about genetic engineering), the tourist trade (which packages 'natural landscapes' for adventure travellers, among other things), the state (which regulates interactions with, and uses of, nature – both human and non-human), businesses (like biotechnology firms) and the non-governmental sector (which includes charities and NGOs like Gene Watch and Earth First!). Together, these knowledge-producing domains condition our collective understanding of nature. Some of them produce knowledges of nature that are tacit, vernacular and seemingly uncontroversial. Others, by contrast, produce knowledges that are sophisticated, technical and challenging. These knowledges circulate between different audiences with different effects. Their influence cannot be 'read-off' a priori.

This book surveys the understandings of nature produced by geographers past and present. It focuses on English-speaking geography because I know it better than geography in continental Europe and the non-Western world. It treats the knowledges of nature anglophone geographers produce as active interventions rather than passive mirrorings of nature's truths. Knowledge of nature is not the same as the 'natural world' it purports to represent. While such knowledge is *about* that world it is not synonymous with it. This knowledge is, if you like, a necessary filter that intervenes between those things we call 'natural' and the way that we, as geographers and citizens, understand and act towards them. This filter allows us to focus on some aspects of those things while casting others into darkness. It also guides our practical interactions with nature, permitting and pro-scribing certain courses of action. As philosopher Ludwig Wittgenstein (1922: 148) famously said, 'The limits of my language are the limits of my world' (Figure 0.1).

Nature considers that particular combination of nature-knowledges produced by anglophone geographers past and present, something that has

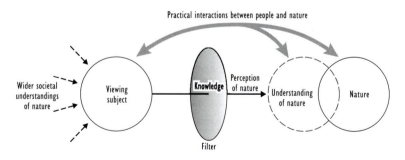

Figure 0.1 Knowledge as a filter structuring our understanding of nature

not been done before. Geography, like most other university subjects today, is filled with specialists who often know very little about what those outside their area of expertise actually think and do. However, because the concept of nature knows no bounds (semantically speaking), one is forced in a book like this to discuss human geography, physical geography and the 'middle ground' occupied by so-called 'environmental geographers'. After all, geographers study everything from floods and plant communities to ideas about people that make reference to their supposed 'natural' characteristics (as racist ideas often do). In short, one is compelled to consider the whole of geography in a book like this one. What such a discussion reveals are the commonalities and fault-lines that give the discipline a structured coherence while, at the same time, threatening any integrity it might claim to possess. That is why any analysis of how nature is understood in geography is necessarily one about the nature of geography.

Second, I have written *Nature* because I think it's important to acknowledge that geography has shaped understandings of one of the most important topics in our lives. Geographers often have an inferiority complex about the wider influence of their thinking. I won't rehearse the reasons for this, simply to note that we risk here *underplaying* the impact of our teaching and research. Quite how important geography is (and has been) in shaping wider understandings of nature is an open question. But I feel sure that the discipline is not the bit-part player it's sometimes seen to be.[1] Generations of students have had their understandings of nature shaped by people like me: that is, supposed 'authorities' in their field. The particular ways that nature has (and, as importantly, has *not*) been depicted in the lecture theatre, the seminar room and in the readings assigned to students has surely had a major impact on how they then view it as citizens, consumers and workers later in their lives. Indeed, the belief that this is the case is one of the motivations for writing this book. Equally, professional geographers have disseminated their research on nature to myriad user-groups beyond university campuses and continue to do so. This is particularly the case in those countries – like the UK – where geography is an established university subject.

THE ARGUMENT

What, then, is my argument in this book? *Nature* is a text whose title is deceptive. It does not do two things that, at first glance, some readers might expect it to do. First, it does not define the word 'nature' and then catalogue how geographers have studied the various bits of nature referred to in the definition. This would make for a very long (and very dull!) book. In effect, it would treat the world as a dictionary where the only interesting issue is how to attach the right words to the right things. Second, *Nature* does not explain the various ways that geographers have studied nature with a view to determining the best or most accurate mode of investigation. Student readers, in particular, will find this disconcerting. Many will suppose that physical geography offers the best route to understanding nature because it is (or aspires to be) a 'science'. What I show in the chapters to follow is that researchers in *all* parts of geography lay claim to understandings of nature that, in their view, tell us something important about it.

This poses a dilemma. Can all these researchers be right, even when their approaches to nature are poles apart? If not, how can we decide between them and eliminate false or erroneous ones? These questions direct us to a hoary philosophical debate between so-called 'realists' and 'relativists'. The latter maintain that absolute truth is impossible since one always understands reality within a particular perceptual or cognitive template. Because templates vary between individuals and groups, relativists maintain that truth is contingent, not given, in nature. In this perspective even science is one world view among others, no better or worse than, say, religion. Realists retort that relativists, absurdly, deny that there is a material world beyond ourselves that we can know accurately. They argue that some ways of understanding the world are more objective than others. In short, they maintain that relativists lead us to the abyss of 'anything goes', where we're forced to accept all perspectives on reality as equally valid. The realism–relativism debate has, at times, been more like a bare-knuckle fight. In the mid-1990s, for instance, the so-called 'science wars' broke out in the United States. Here a group of practising scientists fought back against what they saw as the irresponsible arguments of several sociologists and cultural critics. The latter had argued that scientists construct their knowledge of nature, rather than that knowledge being an accurate reflection of nature's truths. The scientists, understandably dismayed, insisted that science still offers the most secure route to objective understandings of the world (see Gross and Levitt 1994; Holton 1993; Ross 1996). The stand-off between the two groups indicates just how much is at stake in the realism–relativism debate. Those groups in society who can claim that their knowledge of nature is 'objective' or otherwise legitimate can have a lot of power and influence. Equally, those who maintain that objectivity is a fiction concealing the partiality of all perspectives can justifiably challenge truth-claims about the world wherever they hail from.

It is beyond my abilities to resolve the realism–relativism debate. Some commentators have sought to move beyond it altogether and we'll explore their arguments in the penultimate chapter of this book. My tack here is to treat all geographers' knowledge-claims about nature as precisely that: *knowledge-claims* that are each vying for the attention of students, other professional geographers and various constituencies beyond the university. I remain agnostic about the truthfulness or falsity of these claims. Indeed, I could well have subtitled this book '*The Adventures of a Concept*' or '*The Vicissitudes of an Idea*'. These ideas are produced by different researchers and research communities across geography as a whole. They are defined as much by what they exclude as by what they include. Even those ideas that are not formally about nature help us to understand what it is by specifying what is taken to be *non*-natural. The ideas of nature discussed in this book undoubtedly refer to a 'real world' of plants, animals, insects, bodies, ecosystems and much else besides. But since it is no easy matter to decide on which of these ideas is 'better' than the others I give each of them an equal hearing in the chapters that follow. Despite appearances, this doesn't make me a relativist. What I aim to do is show readers how different conceptions of nature are derived in different ways, and mandate different actions on and towards those things we categorise as natural. This includes the 'post-natural' conceptions I examine in Chapter 6 which, by denying there is such a thing called nature, are part of the ongoing tussle to define what it is and what to do with it (or to it). I leave it to readers to judge the merits of the ideas about nature I examine in this book. My aim is to explain them as clearly as I can and to show the practical and moral consequences that follow from accepting or rejecting these ideas. If there's any overarching message then it is this: because knowledges of nature are not reducible to the 'real' nature they depict, it is essential to ask what authorises these knowledges and what sorts of realities they aim to engender.

THE AUDIENCE

At this point, some student readers may be disinclined to read any further. They may be disappointed that this book is about ideas of nature rather than nature itself. They may have picked up this book hoping for a 'no-nonsense introduction' to how nature works from a geographer's perspective. If you're one of these readers I urge you not to give up at the first hurdle. This book will challenge the way you think about nature, of that I'm pretty sure. Part of the challenge is to recognise that ideas about nature are as important as the realities they purport to describe and explain. There is no way to understand nature except through the particular filters and templates that are bequeathed to us by all the knowledge-producing organisations of modern societies. Our experience of nature is rarely direct. Rather, it is thoroughly mediated for us. The filters and templates that mediate our experiences of nature are never neutral. Nor are they passive in relation to that which they refer. By telling us that nature is *this* rather than *that* they govern our understanding of the natural world and how we behave towards it. In the words of one critic, ideas of nature 'are not only products but producers, capable of decisively altering the very forces that brought them into being' (Greenblatt 1991: 6). In a sense, my aim here is to subvert that old schoolyard rhyme 'Sticks and stones will break my bones but names will never hurt me'. To my mind, ideas about nature have a materiality every bit as real as the living and inanimate things those ideas represent.

The sceptical student reader might respond that the role of academic disciplines, like geography, is to expose false ideas about nature circulating in society and replace them with correct ones. In this view, geography stands above the fray. Its role is to produce careful, well-researched understandings of nature free from bias, distortion or prejudice. My own view is that geography, like other disciplines, is never above but always *part of* the ongoing process whereby nature is discussed, debated, used, altered and destroyed. Though geographers often do claim to 'know best' when it comes to the elements of nature they study, I want readers of this book to consider an alternative possibility: namely, that geographers' ideas about nature are part of a high-stakes contest whereby multiple knowledge-producers within and beyond universities struggle to have their views on nature heard. Indeed, the commonplace idea that academic disciplines produce especially trustworthy knowledge can be seen, in part, as a ploy to persuade others in society that academics are worth listening to.

I don't intend to sound cynical. My point, simply put, is that student readers should not place academic disciplines on a pedestal. Geography is characterised by often divergent understandings of nature. These ideas deserve to be understood in their own right rather than as innocent conduits of meaning. The questions I want to ask about them are: Who proposes them? How do they represent nature? What do they exclude? and What consequences follow from these ideas? In a sense, I'm encouraging any sceptical student readers to regard a book about *ideas* of nature as, in effect, a book *about nature* in so far as those ideas frame our understanding of the 'real thing'. These ideas are not reducible to phenomena they refer to; in a sense, they have a life of their own.

Having spoken directly to one of my intended readerships (students), let me now speak directly to another (professional geographers and academics in cognate fields). This readership will have an intimate understanding of much of the material in this book. Trying to say something new to it is much more difficult than it is to students. I have tried to inject some originality into *Nature*. The book's originality (if it has any at all) derives not just from the breadth of material I discuss, but also from the way I organise and interpret it. Synthesising a wealth of geographical research not usually considered in the same intellectual space, I believe I've shed some new light on the nature that geographers study and, as a corollary, on the nature of geography itself.

In practical terms, the only way I can address two audiences simultaneously is to speak to one explicitly and the other implicitly. Academic readers will, I hope, have little difficulty identifying those places where I'm trying to 'add value' to current understandings of how geographers have studied nature and how this has influenced the nature of geography. But in both the prose style and the level at which material is pitched, *Nature* speaks to student readers directly. This book, as befits the series in which it appears, is an advanced introduction. I'm confident that students will find much that is unexpected, exciting and disconcerting in these pages. I'm hopeful too that teachers will want to use the book in modules on nature and environment, on the one side, and those on the nature of geography on the other.

This said, it is symptomatic of how fractured geography teaching is that a book like *Nature*, covering the span of the discipline, will not map well onto many geographers' syllabi. The problem is heightened here because I discuss material not normally considered under the rubric of 'nature'. For instance, I discuss human geography research on social identity and the human body in these pages – research that's a far cry from what physical geographers do. Why, it might be asked, do I roam so widely? There are two reasons. First, the idea of nature is, I argue, a far more pervasive presence in their discipline than geographers have been willing or able to recognise.

It describes more than just the environment (or non-human world). Second, to reiterate a point made in passing above, a good deal of human geography research on what is *not* natural has been important in disciplinary understandings of what, apparently, *is* natural. Some readers might feel that I've cast my net too wide. But my hope is that I show them why their current thinking about the nature of geography – and the nature geographers study – is, perhaps, more limited than they know.

HOW TO USE THIS BOOK

I hope this book will be used rather than simply read. I have written *Nature* in such a way that it could be a core text in geography degree modules on the subject. For students and their teachers I have included features in the book that can sustain a course of study on the topic of nature and the discipline of geography. First, each chapter (except the second) contains Activities. These aim to get student readers reflecting actively on the material presented in the book. Second, each chapter ends with a set of practical Exercises for students. These are designed to reinforce the arguments made and to encourage some free thinking. Third, the Further Reading sections at the end of each chapter have been carefully compiled. In effect, they are a set of reading lists that can be recommended to students as they study up on the arguments of each chapter. Finally, I have included a set of essay questions at the end of the book that can be used in exams and as term-paper assignments. To know quite what to read in order to answer these questions, students will need the guidance of faculty who can refer to the bibliography of this book as well as the recommended readings.

Since most of *Nature's* chapters are quite long I suggest that those who wish to use the book as a core text read them in two or three sittings, depending. For instance, Chapter 3 has three main parts that can be read separately in conjunction with the several references mentioned both in the text and in the Further Reading section. Tutors might like to know that I structure a third-year undergraduate module around the book that comprises thirteen two-hour classes plus a revision class. Prior to each class the students are asked to read a part of each chapter (excepting this Preface, which is read whole prior to the introductory class, as well as the final two chapters which are also read whole). Chapter 1 is read in two parts, Chapters 2 and 3 in three parts and Chapter 4 in two parts.

THE STRUCTURE

Finally, a word on the book's structure. In a short text like this I cannot be comprehensive or exhaustive. Nor do I want to dwell on geography's past at the expense of present-day thinking about nature. *Nature*'s five main chapters (the sixth is a short conclusion) offer an overarching framework into which a mass of material can be slotted – far more material than I can consider in these pages. The material I do include is intended to give students a representative sense of how geographers have understood nature. After a scene-setting Chapter 1, Chapter 2 offers a potted history of how geographers have studied nature and how, in turn, this has influenced the nature of geography. The rest of the book then focuses on present-day geographical thinking about nature. In Chapter 3 I look at work in human geography where nature has been 'brought back in' to research and teaching over the past decade. This rediscovery of nature has, paradoxically, proceeded by 'de-naturalising' our understanding of it and so in Chapter 3 I explain the idea that nature is a social construction. In Chapter 4 I explore the counterposition: the idea that what we call nature is real and can be known in its own right. Most human geographers, I argue, currently have a very different understanding of nature to that of physical geographers and many environmental geographers. Physical geographers, in particular, wear their realist credentials on their sleeves and so in Chapter 4 I explore the grounds for this realist credo in its epistemological and ontological aspects. This human–physical difference impinges on the question of whether geography is a 'divided discipline' and I also explore that question in this chapter. In Chapter 5 I then look at exciting new 'post-natural' thinking (much of it by human geographers) which challenges the society–nature dualism that has long organised disciplinary understandings of nature. In the conclusion I summarise my overall argument and invite students to reflect critically on the politics of their education, whether it relates to nature or any other topic.

Inevitably, a short book on such a large topic reflects my own intellectual predilections. For instance, as a human geographer, my physical geography peers will certainly find Chapter 4 wanting. I therefore apologise at the outset for the biases of argument, simplifications and absences that follow. If I ever have the good fortune to write a second edition of *Nature* I can, perhaps, make amends.

1

STRANGE NATURES

'To dictate definition is to wield . . . power'.

(Livingstone 1992: 312)

TALES OF NATURE

This is not the first book about nature and it will not be the last. To write such a book is, in one sense, to write about everything. After all, one familiar definition of nature is 'the entire physical world' (Habgood 2002: 4). Nature is an all-pervasive aspect of our lives. In fact, it's difficult to think of anything else that's as promiscuously evident in all that we think, say and do. Where previous writers have waxed philosophical about nature it seems to me better to approach it through the concrete forms in which it's routinely experienced and discussed. The following vignettes remind us just how central nature is to our everyday thought and practice – whoever we are and wherever we are.

Blood-ties[1]

In mid-2003 a 13-year-old English boy took his own father to court. 'Daniel' (his real name can't be disclosed for legal reasons) questioned the biological link tying him to his supposed father. Born in 1988 as a result of in-vitro fertilisation (IVF), he spent alternate weekends with his dad subsequent to his parent's divorce when he was just three. But a court-ordered test of Daniel's biological patrimony revealed that a mix-up occurred with his mother's original IVF treatment. The man who was supposed to be his biological father turned out to share no chromosomes with Daniel. His mother's eggs, it transpires, were accidentally fertilised with the sperm of another man. On this basis, a judge ruled that Daniel need never again spend time with the person who, for thirteen years, acted as his father. What has this got to do with nature? In Daniel's case, the lack of a biological link between a father and son was used to terminate a thirteen-year social relationship between a boy and a man. As his mother explained, 'The older he grew the less he looked like or behaved like his so-called father . . . The damage done to that . . . boy is unfathomable' (the Guardian, 23 August 2003). What's interesting here is the suggestion that the absence of a natural (that is, biological) connection has been fundamentally damaging to Daniel's well-being. In effect, his mother argued that this tie alone is more important than the years of time, love and emotional energy that her former husband invested in her son.

Britain's rainforest[2]

Nature can appear in the most unlikely places. Who would've thought that a derelict oil terminal could be one of the most biodiverse sites in Western Europe? In May 2003 an abandoned Occidental facility on Canvey Island, in southern England, was found to contain numerous plant and insect species – many of them endangered and some of them thought to be extinct. These included the shrill carder bee, the emerald damsel fly and the weevil hunting wasp, as well as familiar fauna like badgers and skylarks. Overall, the 100-hectare Occidental site is home to some 1,300 species. But it is threatened with redevelopment as part of the UK government's Thames Gateway expansion plan for nearby London. Intriguingly, nature has returned to this former industrial site because of, not despite, human influence. Some years ago, Occidental dredged thousands of tonnes of silt from the Thames estuary and dumped it over former fields and marshes. It did so to provide foundations for a proposed expansion of the oil terminal that did not, in the end, occur. Then, when the site was abandoned in the early 1970s, it was frequented by children (who played on the site and lit fires) and by bikers (who created trails). The result has been constant disturbance of the plant life growing on the site's fertile soils.

In particular, trees have been unable to take root and this has allowed grasses, wildflowers and shrubs to prosper. In turn, this mixed, low-level vegetation has created the niches that allow the 1,000 animal and insect species identified on the site to flourish. In the words of Matt Shadlow, who runs an invertebrate conservation trust called Buglife: 'This is nature down and dirty'. The dilemma the British government faces is whether to sanction redevelopment of the site (in the interests of an overcrowded London looking for overspill locations) or whether to protect it for its unique ecological qualities.

Sex, violence and biology[3]

Rape is one the most heinous crimes imaginable. It is a crime perpetrated almost exclusively by men almost exclusively against women. It is, according to the evolutionary psychologist Randy Thornhill and anthropologist Craig Palmer, a natural act. In their controversial book *A Natural History of Rape*, Thornhill and Palmer (2000) argue that men rape women in order to spread their genes. The authors see rape as an evolutionary adaptation that lives on even today. According to them, this natural impulse is programmed into males as a reproductive strategy. This is why they subtitle their book 'The Biological Bases of Coercion'. Rapists, they argue, don't normally use excessive force because this reduces the chances that their victims will become pregnant. Unsurprisingly, *A Natural History of Rape* has attracted a torrent of criticism. For instance, the left-wing commentator Kenan Malik has severely questioned the idea that rape is biological. For him, this idea implies that there are limits to how far the incidence of rape can be reduced. After all, if it is 'natural' there is not much one can do about it. More disturbingly, the idea that the male impulse to rape is biological can license a view that it is acceptable because it's 'part of the natural order'. Malik argues strongly that while rape may result from physical urges, it is not reducible to them. For him, rapists *choose* to assault their victims and this is not because of ungovernable biological imperatives but because of life experiences that have influenced their *attitudes* towards women and sex.

Biotechnology's 'new' and 'old' natures[4]

It is one of the paradoxes of modern biotechnology that it can resurrect lost species even as its physically reconstitutes others so that they become 'supernatural'. Consider the following. A team of Italian scientists recently revealed Promethea, the first horse ever to be cloned. This Halfliger foal could presage a new generation of champion racehorses and show horses. Derived from a cell taken from the mare that was its mother, Promethea shows that it is possible to replicate champion horses without the need for stallions, sexual intercourse or even artificial insemination. The owners of thoroughbreds strictly control which other horses their animals copulate with. 'Unnatural' horses like Promethea show that it's now possible to reproduce genetically similar animals generation after generation. Meanwhile, biotechnologists elsewhere are trying to 'bring back the dead'. Cloning – the procedure that made Promethea – can, it is hoped, also be used to resurrect extinct or threatened species, like woolly mammoths, dodos and bucardos. The last of these is a species of goat found in the Pyrenean mountains, the last of which died of natural causes in January 2000. Advanced Cell Technology (ACT) – a Massachusetts-based company – plans to bring bucardos back using cloning techniques. The long-term aim is to create Noah's Arks of frozen genetic material so that any and all threatened animal species can be recreated if the need arises. Not surprisingly, many people are uneasy about these attempts to both supercede and resurrect nature. What, they ask, are the moral implications of biotechnologists 'playing God'? Is it right to 'tamper' with nature in the ways that Promethea's creators and ACT want to do?

Do fish have rights?[5]

In spring 2001 the Texan angling community became the butt of a morally loaded joke. The organisation PETA (People for the Ethical Treatment of Animals) threatened to dose a fresh-water fishing retreat, Lake Palestine, with tranquilliser. Why did it do so? In order to put the lake's fish to sleep so that they would not be caught during the Red Man Cowboy Sporting Division Angling Tournament! The tournament had been scheduled for April Fool's Day and a cadre of park rangers were deployed to prevent PETA seeding the lake with sleeping pills. Sure enough, both the rangers and the anglers were made to look foolish. The tournament went ahead as planned, and it's not difficult to understand why. A few litres of tranquilliser, however potent, could not have much effect on a 40-billion-gallon lake containing countless fish! But the PETA joke was a serious one. It challenged the received view that fish suffer no pain when their mouths are snagged with sharp hooks and then reeled in. In short, it questioned the image of fishing as harmless 'recreational sport'. For some, of course, the suggestion that animals – including fish – may have rights is ludicrous. But PETA is by no means alone in its arguments. The year before the Lake Palestine incident a distinguished Harvard law professor and Boston attorney, Stephen Wise, published *Rattling the Cage*. This erudite and rigorously argued book demonstrated the arbitrariness of limiting legal rights to humans alone. In effect, Wise gave legal substance to the aspirations of PETA and other animal-rights organisations, as well as the ideas of famous animal-rights philosophers like Peter Singer. His book aims to change Western mindsets about non-human species. Just as we now see the medieval practice of witch-burning as barbaric, so Wise and others hope to persuade us that we are wrong to unthinkingly use animals as means for our own ends.

Crisis, what crisis?[6]

Patagonian toothfish can live up to fifty years and take about ten to reach sexual maturity. They are highly valued in restaurants in Japan and the USA among other countries. They are currently subject to heavy overfishing, most of which is illegal. In mid-2003 a dramatic instance of this was broadcast worldwide. News channels showed the *Viarsa* (a Uruguayan fishing vessel) being pursued for some two weeks by the *Southern Supporter* (an Australian customs ship). The *Viarsa* had taken toothfish from Australian territorial waters without permission and in breach of regulations. Meanwhile, in the northern hemisphere, the Aral Sea has shrunk to less than half of its 1950 volume and area. Irrigation measures instigated by the former USSR have robbed the sea's feeder rivers of water. It is now little more than a saline desert. To make matters worse Kazakhstan proposes to dam the northern part of the sea, leaving the south part with only a fraction of the 1000 cubic kilometres of water inflow per year needed to maintain existing shorelines. Further north still, the St Roch – an ice-breaker belonging to the Royal Canadian Mounted Police – became the first-ever ship to complete the fabled Northern Passage from west to east in September 2001. So thin and fragmented had the Arctic ice packs become that the St Roch could cut through from the Bering Sea to Greenland via Banks Island. Finally, in many countries worldwide there's a growing realisation that plants formerly classed as 'weeds' or else as valueless foliage might, in fact, be enormously useful. Precisely because they've been seen as 'useless' in the past they are often close to extinction today. Bogbean, yellow gentian and panax ginseng are just a few of the wild plant species now thought to have medicinal properties hitherto unappreciated. Consequently, some are making frantic attempts to conserve them before it's too late. What do these four cases have in common? The answer is that they're all about the destruction of nature. For some environmentalists ours is an era of 'environmental crisis', one where we're witnessing 'the end of nature' (McKibben 1990). In fact, it's become a commonplace to hear the word 'crisis' uttered in relation to humanity's current usage of natural resources. But not everyone agrees that we're in the grip of a crisis. In 2001, for instance, a Danish statistician called Bjorn Lomborg published a controversial book entitled *The Skeptical Environmentalist*. Lomborg produced a mass of evidence to show that humanity's treatment of the environment is, in his view, *improving*. What's more, he criticised environmentalists for scaremongering and for exaggerating the scale of environmental problems. Not surprisingly, green activists have attacked him mercilessly and he was even accused of manufacturing and falsifying much the evidence used in this book.

Having fewer genes is good for you[7]

The Human Genome Project – an internationally funded attempt to describe humans' genetic make-up – has revealed that *Homo sapiens* have fewer genes than expected. In 2001, an initial analysis of the human genome revealed that we are comprised of some 30,000–40,000 genes – only two-thirds more than a fruit fly. This raises the question of how humans can be so far in advance of other living species with so few extra genes. The answer, according to those who believe that genes do not explain much about people's physical and mental abilities, is the social and cultural milieux in which genetic capacities are expressed. Those who favour 'nurture' as an explanation of human behaviour over 'nature' insist that humans' biological capacities are highly conditioned by societal factors. This view challenges 'genetic determinists' like the right-wing American writer John Entine. His mould-breaking book *Taboo: Why Black Athletes Dominate Sports and Why We're Afraid to Talk About It* (2000) argues that the preponderance of African-Americans excelling in professional sport in the United States must be down to DNA. Against this, those who look to societal factors would argue that professional sports is one of the few available routes out of poverty for many African Americans.

These seven stories about nature are interesting and memorable. If most of them seem unusual, it's only because most of us rarely pause to consider how deeply insinuated into our thought and practice nature is. The stories above recount only a few of the countless ways in which nature is made manifest in numerous walks of life worldwide. But what do these vignettes actually tell us about the subject of this book? In other words, what lessons about nature can we draw from our seven rather different stories?

ACTIVITY 1.1

On the basis of the seven nature tales above, answer the following questions:

- What is nature?
- Where is nature?

Don't rush. This activity will take you some time. Read the stories again slowly and think hard about these two questions. Try to stand back from the details of the stories. See if you can identify some broad similarities and differences between them that will allow you to answer the two questions posed. Answering these questions carefully will help you understand what this book is about.

The two questions in the activity above are intended to get you thinking about the above-mentioned pervasiveness of nature in contemporary life. Nature is one of those topics that, if you reflect on it for just a second, seems to pop up in all manner of different contexts – from discussions of human genes to those about fish and cloned mammals. But the two questions serve a second purpose too. When answered in relation to the seven stories recounted, they are also designed to challenge our habitual ways of thinking about nature. Let me explain.

One common definition of nature is that it is the non-human world. According to this definition, the word 'nature' is more or less synonymous with the word 'environment'. In our seven stories, this definition would encompass everything from toothfish to bucardos to shrill carder bees. Even without having to formally describe any of these things as 'natural', it is implicit that this master category encompasses them according to conventional usage. But these stories also remind us that nature means 'the essence of something' as well. Using this second, broader definition we see that nature also encompasses humans too – whether it's the genetic traits of African-American sportspeople or the blood ties between parent and child. Thus, to utter a phrase such as 'It's in their nature' is to say that a person has certain physiological or psychological qualities that help to make them the kind of person they are. This links to an even broader conception of nature as the inherent force ordering both humans and non-humans. To take an example from our stories, we see this force referred to when critics of biotechnology argue that it goes 'against nature' by creating things like cloned foals and Pyrenean goats. Likewise, when some environmentalists talk about the non-human world as a self-regulating system (as in the famous 'Gaia hypothesis') they are using the idea of a transcendental God-like power. In this third definition, we might think of nature with a capital N as opposed to its various component parts in the human and non-human worlds. Table 1.1 summarises these three principal definitions of nature.

Similarly, if we ask 'Where is nature?', our stories remind us just how limited conventional understandings are. These understandings take two related forms. One sees nature as primarily located in the countryside,

Table 1.1 The meanings of nature

	The non-human world	The essence of something	An inherent force
The environment/ external nature	✓	✓	✓
Humans		✓	✓

in rural areas and in wilderness zones. The other, more specifically, locates nature by its visible types (forests, mountain ranges, water bodies, deserts etc.). In both cases, the first definition of nature (the non-human world or environment) is rendered geographical by placing it outside the domain of human settlement. But our seven nature tales challenge this conventional way of thinking about where nature is. Occidental's former Canvey Island site, for example, reminds us that what we call 'nature' is very much on all of our doorsteps – even if we live in large, densely populated cities. Meanwhile, the debates over *A Natural History of Rape* and *Taboo* recall the unavoidable fact that we humans are natural animals too. At some level, our biological capacities condition what we are able to do at all stages of our lives. In this sense, nature is always already *here* – intimately a part of us – not just somewhere else or beyond us.

To summarise, it's clear that nature knows few bounds when one considers the range of contexts in which we encounter phenomena that, whatever their apparent differences, we classify as 'natural'. In various forms, it appears in everything from biotechnologists' labs to brownfield sites. And it encompasses everything from the human body to fish and foals. Given this, it may be wondered how a relatively short book like this can sensibly discuss a topic as colossal as nature. My response is that this is not a book about nature but, rather, about how *geographers have understood nature*. As we shall see, this does not narrow things down as much as might be supposed. But it nonetheless gives a distinct focus to Nature. In this introductory chapter I want to explain why – and how – I choose to discuss nature through a geographer's lenses.

KNOWLEDGES OF NATURE

Geography is one of several subjects devoted to the study of natural phenomena. Nature is not, of course, the only thing that geographers study. But it's long been recognised as a major disciplinary preoccupation. This stretches back to the foundations of Western geography as a university subject in late-nineteenth-century Europe and North America. At this time the now-familiar academic division of labour between the natural sciences, the social sciences and the humanities was starting to take shape. New disciplines like sociology and biology were being created, and what they had in common was that they were relatively *specialised* fields of research and teaching. In this context, those who first championed geography saw it as a uniquely *integrative* (or 'composite') discipline that would synthesise the Balkanised knowledges produced by more analytical subjects. The 'geographical experiment' consisted in trying to bring society and nature 'under the one conceptual umbrella' (Livingstone 1992: 177). In its foundations, then, geography was defined not as the study of nature per se but, rather, as the study of society–nature relationships. It was intended to be the 'bridging' subject that spanned the gaps created by academic specialisation.

Well over a century later, geography is an established university (and school) subject worldwide. Though its reputation varies from country to country, it's widely recognised that geographers study human impacts upon nature (and vice versa). What has changed since the discipline's foundation as an integrative subject, is that (ironically) there has been specialisation within geography itself. This is certainly true of Western geography. Apart from the 'divide' between human and physical geography, both 'sides' of the discipline are split into subfields like economic geography and geomorphology. Those working in the disciplinary 'middle ground' are now relatively few in number and focus, among other things, on natural hazards and natural-resource management (see Figure 1.1). These 'environmental geographers' (as they're sometimes called) also now have to share the study of society–nature relationships with environmental science, earth science and environmental management – three increasingly popular interdisciplinary fields that bridge physical-science perspectives on the environment in the first and second cases, and physical and social-science perspectives in the third case.

I'll say more about which specific aspects of nature geographers study in the next section. For now I simply want to note that geography is just one of several disciplines producing knowledge about nature – it's but a single player in a crowded field. Ask yourselves what the other disciplines

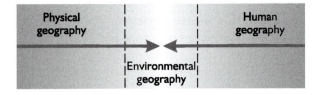

Figure 1.1 Geography's main branches. The nearer the middle, the less 'pure' the human and physical geography become

are. If you think about it carefully you'll recognise that it's not just the physical sciences, like chemistry or metallurgy, nor simply the engineering and materials sciences. Nor is it, additionally, just the medical or sports sciences, which look at the nature of human biology. It is also parts of the social sciences and the humanities. For instance, anthropologists have long examined how indigenous peoples utilise local environments. Meanwhile, environmental historians have studied previous human impacts upon the natural world (and vice versa) going back tens and hundreds of years. In other words, the whole spectrum of academic subjects produce a range of knowledges about nature (see Box 1.1).

Box 1.1 ACADEMIA AND THE STUDY OF NATURE

A simple, but effective, way to get a feel for how widespread academic interest in nature is is to list some titles of books published in different disciplines. In the titles below you should be able to roughly guess the discipline and roughly understand what 'nature' means in each case:

- *Ideas of Human Nature: An Historical Introduction* (Trigg 1988)
- *Essentially Speaking: Feminism, Nature and Difference* (Fuss 1989)
- *The Rights of Nature: A History of Environmental Ethics* (Nash 1989)
- *Physical Geography: Its Nature and Methods* (Haines-Young and Petch 1984)
- *Against Nature: Essays on History, Sexuality and Identity* (Weeks 1991)
- *Not in Our Genes: Biology, Ideology and Human Nature* (Rose and Lewontin 1990)
- *The Nature of the Environment: An Advanced Physical Geography* (Goudie 1984)
- *What is Nature?* (Soper 1995)
- *The Scientific Nature of Geomorphology* (Rhoads and Thorn 1996)
- *The Nature of Weathering* (Yatsu 1988).

But things don't end there. Academic disciplines are only one of several domains where knowledges of nature are produced. If you think again about the seven vignettes with which I began this chapter it's not hard to list the variety of institutions, organisations and professions that have something to say about nature. As I noted in the Preface, these include newspapers, movies, television programmes, popular books, businesses, governments, courts, charities, and independent think tanks. Aside from academics, there are pundits, broadcasters, freelance writers, environmental activists and lawyers (among others) who routinely consider nature in their discourses. Together they produce a constant stream of information not only about what nature is but about 'appropriate' ways to use it, control it or alter it. On any given day of the week a veritable mountain of knowledge about nature is circulated, communicated and disseminated within and between societies worldwide. As individuals we are all exposed to particular mixtures of nature-knowledge over our lifetimes. Our understanding of nature is, obviously, heavily influenced by the 'truths' and 'norms' about nature imparted to us through the variety of knowledge-producers mentioned above (Box 1.2).

Box 1.2 KNOWLEDGES OF NATURE

At one level we are all producers and consumers of knowledges of nature. But what is 'knowledge'? Knowledge is sometimes defined as distinct from 'opinion', but in this book I take a broader view. Knowledge is any form of understanding that can be articulated verbally, textually or pictorially. In other words, knowledge is how we represent the world in which we live to both ourselves and to others. Knowledge is acquired through observation, interaction with other people and engagement with the material world. It is capable of being modified over time and space, either slowly or more quickly depending on the circumstances. Typically, knowledge exists as more or less established bodies of knowledge that distinct groups of people share in common. Equipped with this broad definition, we can make some useful distinctions that help us get a better handle on the character of knowledge. To begin with, all

---

knowledge has a point (or points) of origin, a referent (or referents) and an addressee (or addressees). The first describes the institutions, groups or individuals who promulgate a particular body of knowledge or specific knowledge-claims. The referent/s of knowledge are those particular things referred to in any knowledge-claim or body of knowledge. Referents can be either material things or other bodies of knowledge. The addressees of knowledge are the intended audience for a particular representation or set of representations of the world; in effect, they are the 'consumers' of knowledge. The trio of origin/s, referent/s and addressee/s help us distinguish one body of knowledge from another. In the second place, we can distinguish tacit (or taken-for-granted) knowledge from formal (or codified) knowledge. The former is all that knowledge that's so deeply internalised that it's simply 'common sense'. Though this knowledge is capable of being articulated formally it is rarely necessary to spell it out. The latter is all that knowledge that's explicitly articulated, either because of its complexity, its novelty or its specialised character. The distinction between tacit and formal knowledge is closely linked to that between lay (or vernacular) knowledge and expert (or technical) knowledge. The former is 'ordinary' knowledge that we all deploy in everyday life. The latter is higher-level knowledge used for specific purposes and intended for specific audiences. Technical knowledges are often characterised by their exclusivity in terms of who produces them and who consumes them.

I mention all this for two reasons. First, it is important to appreciate that the understandings of nature produced by geographers and other academics must, in one sense, *compete* among themselves and in relation to those nature-knowledges by myriad non-academic organisations (Figure 1.2). This may strike student readers as a peculiar claim at first sight. After all, you might think that academic knowledge of nature is relatively objective whereas that produced by other organisations is often less so. Indeed, some of this non-academic knowledge is patently fictional – think, for instance, of the supernatural world depicted in the movie *Jurassic Park*.

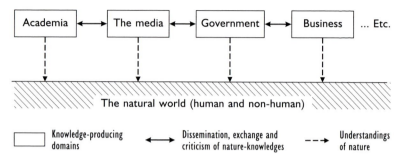

Figure 1.2 Knowledges of nature

On this basis, it might be thought that academic understandings of nature are 'special' because they aim for objectivity and are rigorously arrived at. In this view, then, what non-academic organisations normally do is simply use and report the 'facts' or 'truths' about nature disclosed by academic researchers. The suggestion here is that these organisations are but 'relays', bowing down before the altar of academic understandings of the natural world.

But life is not that simple. To begin with, researchers in the same or different academic disciplines may fundamentally disagree in their analyses of phenomena that are deemed wholly or partly natural. Global warming is a current case in point. Some atmospheric scientists still resist the idea that a natural 'greenhouse effect' has been exacerbated by the under-regulated emission of airborne pollutants. Still others disagree profoundly about the likely effects of global warming. Academic disputes aside, non-academic organisations often have their own agendas or are forced to simplify the often-complex information academic researchers produce. An infamous case is the research of Arpad Pusztai. Pusztai made headlines in 1998 because of his research into the effects of genetically modified foods. In that year a British current-affairs programme (*World in Action*) reported Pusztai's unpublished study showing that rats consuming GM potatoes experienced unusual physiological changes. Unsurprisingly, the many environmental groups who were anti-GM foods seized upon Pusztai's findings, while the pro-GM lobby sought to undermine his methods and even his character (see Box 1.3). In short, Pusztai's research became a battlefield, with different interest groups selectively depicting his findings. Finally, it's worth remembering that the legitimacy of 'expert knowledge' has been publicly called into question in recent years. To

Box 1.3 THE 'PUSZTAI AFFAIR'

In 1998, the safety of genetically modified (GM) foods became a national preoccupation in the UK because of the research of Arpad Pusztai. Pusztai's research team, based at the Rowett Institute in Scotland, fed GM potatoes to rats in an attempt to determine what physiological effects the potatoes have when compared to non-GM varieties. When the results of the trial were eventually published – in the respected medical journal the *Lancet* in 1999 – they showed that there were differences in the gut linings and organ sizes of the two sets of rats used. For some, these differences implied that GM foods might cause health problems. But these health concerns had already been expressed in a more informal way in 1998 by Pusztai himself, first in an interview for the news programme *Newsnight* (shown on BBC Two) and later for the current affairs show *World in Action*. In these TV appearances Pusztai expressed his concern that GM foods might not be safe for human consumption and he speculated that biotechnology firms were using people as guinea-pigs for untested foodstuffs. This led to his employer firing him, to an outcry from biotechnology firms, and to members of the UK scientific establishment (like the government's chief scientist Sir Robert May) questioning Pusztai's integrity. The controversy centred on the quality of Pusztai's research and the wider conclusions that could be drawn from this one investigation into GM foods. Those with a vested interest in GM foods sought to downplay the veracity of Pusztai's findings, while those suspicious of their benefits championed Pusztai's research. Because of the extensive media attention that the Pusztai affair received, the British public became very concerned about the safety of GM foods. The UK government was ultimately forced to acknowledge the need for proper scientific studies of the health and environmental impacts of GM foods. Meanwhile, Pusztai's reputation as a scientist has been damaged by the events of 1998/9. He will be remembered less for the quality of his research and more for his courage as a whistle-blower or else for his malign influence on the British public's perceptions of GM foods.

take a British case once more, the early 1990s saw the outbreak of BSE in cattle, an infectious neuro-degenerative disease that has a human form (Creutzfeldt-Jakobs Disease). By 1993 some 100,000 farm animals were affected. Animal biologists working in universities and for the British government were fairly confident about the causes of, and remedies for, these two diseases. They were equally confident that BSE-infected meat posed no public health risk, although measures were taken to remove such meat from the food chain. Drawing upon the authority of science, the British government devised a policy to manage and eliminate BSE. By 1996 it turned out that this policy may have been founded on mistaken beliefs. Hitherto ignored or marginalised scientific research suggested that as many as half a million people could have CJD because of eating meat from BSE-infected cattle. The British governments' attempt to legitimise its policy decisions by depicting scientific knowledge as 'reliable' and 'true' backfired. The result was not just a collapse in the British cattle industry, but also a public loss of faith in the ability of professional researchers to deal in certainties rather than merely supposition and conjecture.[8]

This brings me to the second reason for emphasising the multiple domains in which knowledges of nature are produced. *It's important not to confuse knowledges of nature with the 'natural' things those knowledges are about.* The relationship between knowledge and the world it depicts has preoccupied generations of philosophers. For now, we simply need to recognise that without knowledges of nature we can never really come to know the nature to which those knowledges refer. This is not say that we only comprehend nature by means of formal statements about, and mental understandings of, it. Touch, sound and smell matter immensely too. But it remains the case that we use tacit and explicit knowledges to organise our engagements with those phenomena we classify as 'natural'. There is, in short, no unmediated access to the natural world free from frameworks of understanding. These frameworks organise the way that individuals and groups view nature and delimit where the natural ends and the unnatural, non-natural or artificial begins.

Some readers might object that many understandings of the natural world are relatively direct and unmediated – untroubled by any 'detour' through inherited frameworks of understanding. For example: they might cite the farmer whose intimate knowledge of soils and crops comes from years of practical experience. Equally, they might cite the child whose growing understanding of what their body can do comes, in part, from physical play in gardens and parks. Yet, persuasive as these examples might seem, I'd make two counter-arguments. One is that even the practising farmer and the developing child learn about nature, in some measure, through knowledge passed down to them by their families, parents or what have you. The other is that much of the nature we think we know about has, in fact, rarely been seen, touched, heard or experienced by us. Rather, we've been told what it's 'really like' by all those knowledge-producing institutions and professions mentioned above. For instance, personally I've never visited a glacier. How, then, do I know what a glacier is and how it moves? The answer is that I only know through a mixture of my geographical education and the occasional television documentary. I simply take it on trust that the knowledge fed to me is a fair representation of what 'real glaciers' are like. My understanding of glaciers is, in other words, derived from 'second-hand non-experience'.

In a broad sense, then, knowledges that are inherited, assimilated and learned act as a filter that mediates between ourselves and nature – whether the nature in question be the non-human world or our own bodily natures (see Figure 0.1 again). However, it's worth noting that knowledges of nature (and indeed all knowledges) come in three forms. *Cognitive knowledges* make claims about what is (and is not) natural; they seek to describe and explain those things we call 'nature'. *Moral* (or *ethical*) *knowledges*, as the name suggests, entail value judgements about the propriety of what is (and, again, is not) done to those things we consider to be natural. Finally, *aesthetic knowledges* seek to instruct us on what is beautiful, uplifting or otherwise pleasurable about what we call 'nature'. Aesthetic knowledges are less about what is 'good', 'right' and 'just' (this trio is the domain of moral knowledges) and more about what is edifying and sensually satisfying. It's important to note that moral and aesthetic knowledges come in two forms: *descriptive knowledges* and *normative knowledges*. The former are currently existing moral and aesthetic knowledges that have some purchase in society. The latter are suggestions about the kind of moral and aesthetic knowledges we *should* adhere to in the future. Normative knowledges are usually critical of descriptive knowledges and the practices they licence. We should also note that moral knowledges (and to a lesser extent aesthetic ones) are sometimes 'read-off' by people from cognitive claims. For instance, in the case of the story about 'Daniel', he and his mother deemed his father's claims to parenthood 'illegitimate' because of the lack of a blood tie. All this is summarised in Table 1.2.

Table 1.2 Types of knowledge

	Cognitive	Moral	Aesthetic
Descriptive	✓	✓	✓
Normative		✓	✓

Finally, before I consider what all this has to do with geography specifically, let me say something about what is sometimes called the 'materiality of knowledge'. One common view of knowledge is that it's less 'solid' and less influential than the 'hard stuff' of the world, like bricks and mortar. Knowledge is often seen as being 'immaterial' when compared to a physical world usually seen as more 'real' and tangible. Against this, I would argue that knowledge is as material as the things to which it refers. This is not to say it is the *same* as the those things; if it were it would lose its relatively autonomy. A particular body of knowledge can have tangible effects to the extent that people believe it to be legitimate, truthful and valid. One only has to look at the history of ideas to know that this is so. For instance, until the publication of Charles Darwin's *The Origin of Species* it was commonly thought that God created life on earth. Almost 150 years later, the theory of evolution has more or less replaced this theological perspective. I say 'more or less' because even today some people hold fast to the 'creationist' line. In the USA, for instance, the past decade has seen a heated debate between evolutionists and several religious groups who proclaim the divine provenance of all living things. This debate has even made it into the courts, where both sides have sought their legal right to stop the others' views being taught on school syllabi. A legal conflict like this illustrates just how much is at stake when certain knowledges seek legitimacy in the public realm. Knowledges of nature are multiple in their origins, their meanings, their referents and their audiences. Together, they materially shape understandings of, attitudes towards, and practices upon those numerous things we describe as natural things. In short, the contest whereby certain knowledges of nature gain purchase in any society (or some part thereof), while others are marginalised, is a high-stakes one. According to some, it's a contest over which knowledges become *hegemonic* (Box 1.4).

Box 1.4 HEGEMONY

The *Oxford English Dictionary* defines a 'hegemonic' organisation, group or individual as one that 'rules supreme'. However, this does not accurately capture the meaning of the term 'hegemony'. The term is most closely associated with the writings of an Italian Marxist called Antonio Gramsci. Gramsci was incarcerated by the Fascist Italian government between 1928 and 1935. While imprisoned, he reflected on how citizens lend their assent to forms of government that curtail their freedoms and adversely affect various other aspects of their lives. He came to believe that powerful groups in any society get their way not through coercion or force but through persuasion and assent. For Gramsci, hegemony described a process whereby dominant factions of a society portrayed *their* beliefs and values as those good for society as whole. Over time, these hegemonic ideas take hold not just by being repeated endlessly (in the media, in schools, in political speeches etc.) but also by being embodied in policies and institutions. For Gramsci, hegemonic ideas ultimately become 'common sense' for the mass of the populace and this is what makes them so effective as tools of control. As the Marxist cultural critic Raymond Williams observed some years after Gramsci wrote his prison notebooks, hegemony is 'a lived system of meanings and values – constitutive and constituting – which, as they are experienced as practices, appear as reciprocally confirming' (Williams 1977: 110). What has this got to do with knowledges of nature? On the one hand, the answer is 'not a lot'. Many knowledges of nature can hardly be accused as being tools for the control of people or even the non-human world. On the other hand, though, because nature is such an all-pervasive aspect of our collective thought and practice, the way it is understood is manifestly important. Hegemonic ideas about nature are those *general* understandings of human nature and the non-human world that are more or less 'taken for granted' in any society. These ideas have a history, a geography and a sociology to them. In other words, they begin with someone or some organisation, they then spread across space to influence greater numbers of people, and they reflect, in some measure, the agendas of those who promulgate these ideas. For instance, it serves white racists very well to insist that people of colour are 'naturally less intelligent' than Caucasians. Such beliefs gain extra credence when some authority can be invoked to justify them. For instance, in the mid-1990s a controversial American book entitled *The Bell Curve* sought to offer scientific proof that IQ varied according to one's class and, by implication, 'racial group' (Hernstein and Murray 1996).

Of course, the idea that white people are more intelligent than non-whites is not a hegemonic idea in Western countries today, though it arguably was in the past. The reason it is only taken seriously nowadays by a minority of people is because it has been successfully challenged by what were once *counter-hegemonic ideas*. These counter-hegemonic ideas (like the suggestion that non-whites do less well on IQ tests than whites because they often suffer a worse education) become hegemonic once enough people can be persuaded that they are valid. But it would be a mistake to think that counter-hegemonic ideas are always more objective and less 'biased' than the hegemonic ideas they oppose. Arguably, *all* hegemonic ideas reflect the agendas, aims and objectives of those expounding them.

Sources: Gramsci (1995); Johnston et al. (2000); Williams (1977)

NATURE AND GEOGRAPHY

In the previous section I said a lot about knowledges of nature. This followed an introductory section in which we established that nature means more than just 'the physical environment'. We're now in a position to discuss the particular knowledges of nature produced by professional geographers. One starting place is to apply the discussion of different types of knowledge to geography. This done, we can then focus on geographers' knowledges of nature specifically rather than the nature of their knowledge in general.

ACTIVITY 1.2

Read Box 1.2 and also the part of the previous section where cognitive, moral, aesthetic, descriptive and normative knowledges were defined. Once you've done this, answer the following question: which of these knowledges do geographers produce in their research and then disseminate through their teaching and external activities?

What is your answer? It's likely that, using Box 1.2, you think geographers are producers of formal, expert knowledge for a range of addressees, including students, other academics and outside bodies like governments. If this was your answer you'd be correct. There's nothing tacit, for example, about human geographers' theories of uneven development, while a physical geographer's research into gravel-bed rivers is hardly intended for consumption by the general public. What, though, about the five knowledge types discussed in the previous section? Chances are you rightly identified that geography produces a lot of cognitive knowledge. But did you know that geographers also make moral and aesthetic claims, often of a normative sort? For instance, several human geographers write about spatial injustice (as when people in one place suffer lower levels of health-care provision when compared with other places in the same country), and still others have examined our emotional attachments to particular landscapes. So, to summarise, professional geographers produce a wide range of higher-level, formalised knowledges. Simplifying somewhat, we can say that the discipline generates cognitive knowledge for the most part, with human and environmental geographers also producing a fair amount of moral knowledge, and human geographers not a little aesthetic knowledge.

So much for knowledge in general. What about geography and knowledges of *nature*? I introduced the previous section by talking about academic geography's origins as a 'bridging' subject that crossed the Maginot lines dividing specialist understandings of the world. I also observed that, over time, anglophone geography has split into two 'halves' with a shrinking 'middle ground'. I'll say more about the two halves below, but let me begin by discussing this middle ground. It may well be shrinking, but it has by no means disappeared. The 'nature' that environmental geographers study is accurately captured in the name used to describe them: that is, the physical environment (our first definition of nature). But if this was *all* there was to their approach to nature it would be little different to any number of other environmental researchers and teachers. It's important, then, to add that environmental geographers look at the environment (i) in relation to specific human interpretations and uses of it and (ii) in an integrative, interconnected way. In other words, they do not do 'pure environmental research' nor are they normally specialists in the topical sense. Rather, they look at how everything from peoples' perceptions of environmental hazards through to land-use practices through to the physical behaviour of, say, a tornado, *combine* in particular times and places with more or less disastrous or benign consequences. This sort of 'human-environment' research traces a lineage back to George Perkins Marsh's *Man and Nature* (1864) and is continued today by the likes of Bill Adams (a British geographer) and Billie Lee Turner II (an American geographer). Some environmental geographers approach things more from the physical side (like Andrew Goudie, a well-known arid-lands specialist), while others look more at the human dimensions (like Tim O'Riordan and Susan Cutter, who are both interested in environmental management). What they share, though, is a commitment to examining the reciprocal relations between societies and their environments. If environmental geographers are 'specialists' at all it is either regionally (in terms of where they undertake their research) or because they are expert about a particular natural hazard or natural resource. For the most part, these geographers combine a broad intellectual training with a detailed grasp of how social and physical processes intertwine. While most of them look at present-day issues, not a few take a longer historical perspective.

Other geographers, by contrast, prefer to study the physical and human worlds alone in either contemporary or historical contexts. For reasons to be explained in the next chapter, anglophone geography has become something of a 'divided discipline'. The benefits of this are that geographers have been able to specialise rather than be the 'jacks of all trades' that environmental geographers are sometimes seen to be. Physical geography is, these days, comprised of the following subfields: geomorphology, hydrology, climatology (with meteorology) and biogeography (with soils). Quaternary studies is, increasingly, considered to be a fifth specialism (see Figure 1.3). Overall, physical geography is a 'field discipline' or 'earth science' that routinely undertakes 'pure environmental research' and a fair

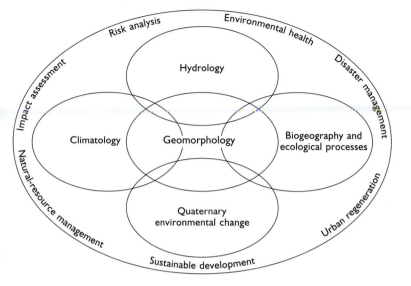

Figure 1.3 The five overlapping subdisciplines of physical geography in the context of multidisciplinary and problem-oriented themes. Reproduced from Gregory et al. (2002)

bit of applied research too. In other words, it looks at 'real environments' past and present, choosing to bracket out the human element and often seeking appropriate ways to manage or modify those environments. By contrast, human geography focuses very much on the world of people. Its main topical branches are economic, social, cultural, development and political geography while, spatially speaking, it has an urban and a rural arm (the latter including agricultural geography). Indeed, it is symptomatic of geography's divided nature that most professional geographers are labelled as physical or human geographers, while geography degree students normally carry similar epithets. If you look at the work of the better-known human and physical geographers, the differences are striking. Take Neil Smith and Olav Slaymaker, two leading geographers of their generations. The former is a Marxist who theorises uneven development, while the latter is a fluvial geomorphologist! 'Never the twain shall meet', as they say.

What's all this got to do with nature? My answer is: more than meets the eye. Superficially, it appears that physical geographers are the discipline's greatest producers of knowledge about nature. After all, they focus on environmental processes and forms and they comprise a larger research community than do environmental geographers. Like environmental geographers, the 'nature' that physical geographers study is very much that denoted by the first meaning of the term: that is, the non-human world. Physical geographers' distinctive contribution to understanding nature is that, unlike the laboratory sciences, they investigate dynamic real-world environments. If you like, they borrow laws, theories and models from the physical sciences (chemistry, physics etc.) and operationalise them in 'live settings'.

So far so good. But the assumption that only physical and environmental geographers study nature, while human geographers focus on other things, is misplaced. To be sure, human geographers do not study the physical environment in any *direct* way. In other words, almost by definition, their research does not seek to comprehend how the non-human world 'really works'. But an increasing number of human geographers are interested in how different sections of any society *interpret* the environment. In other words, they analyse the knowledges of the non-human world that circulate within and between real-world groups and organisations. Take the research of Bruce Braun, for example, my sometime co-editor. His book *The Intemperate Rainforest* (2002) examines the conflicting ways in which the forests of British Columbia have been cognitively, morally and aesthetically 'framed' since the province was colonised by the British in the mid-nineteenth century. His concern is not with the details of forest ecology or, say, the environmental de/merits of clear-cut logging. Rather, he is interested in how the *same* forest lands are interpreted in *different* ways by the several groups who have sought to determine their fate over the decades. His question is: who defines the forest in what ways and for what ends?[9] In a sense, human geographers like Braun produce knowledge about other peoples' knowledges of the environment. This type of research has a long pedigree, even if the theories and methods used to do it have changed considerably. For instance, as far back as 1947, the geographer J.K. Wright discussed the role of imagination in human understandings of the material world.

Ideas, of course, influence peoples' practices – but the latter are not reducible to, or wholly determined by, the former. I mention this because other human geographers have taken a keen interest in how different forms of social organisation lead to different practical engagements with the environment. These geographers examine how certain ways of organising economic, cultural, social or political activities have specific environmental consequences. Unlike environmental geographers, these researchers do not explore those consequences in much geophysical or biochemical detail. Rather, they study the *human practices* that give rise to those consequences. An example is the research of Gavin Bridge, a colleague of mine. Bridge's work on the modern mining and forest-products industries shows how their embedding within a distinctively *capitalist* economic system leads to specific ways of mining and managing timber that give rise to particular forms of environmental damage (e.g. Bridge 2000).

But human geographers' contributions go further than this. If we consider the second definition of nature – namely, the essence of something – it's possible to identify two further ways in which human geographers investigate nature. First, all societies, economies, cultures and polities have an essence as much as physical environments do – if by that term we simply mean a definite 'character' or way of operating. In this broad sense, human geographers investigate the spatial 'nature' of social, economic, cultural and political processes, practices and events. Likewise, physical and environmental geographers also study nature in this second sense, since they are concerned with the 'nature' of environment and the 'nature' of society–environment relationships respectively. But this very general sense of nature as essence, it's readily apparent, potentially takes us a long way from the concerns of this book. In effect, it would make *Nature* a book about *everything* that geographers say and do!

However, a second implication of the definition of nature as 'the essence of something' is more helpful for our understanding of the 'nature' that contemporary human geographers study. This definition, recall, implies that humans – as biological beings – have a nature just as much as the non-human world does. In recent years, human geographers have questioned whether and how such a nature exists. Unlike many researchers in the medical and psychological sciences, several human geographers have shown that what we call 'human nature' (bodily and mental) is not *simply* natural. Books like *Places Through the Body* (Nast and Pile 1998), *Mapping the Subject* (Pile and Thrift 1995) and *Body/Space* (Duncan 1996) have taken issue with the idea that our physical and mental 'natures' are asocial, 'given' and 'fixed'. In effect, they attempt to 'de-essentialise' that which *seems* natural. To de-essentialise is to show that what seems fixed in nature is either changeable or else was never really fixed in the first place. This is contentious because many eminent thinkers – like the neuroscientist Steven Pinker (2002) – believe humans have natures that help explain the kind of people we become. By contrast, a cohort of human geographers argue that ideas about human nature are just that: *ideas* which need to be 'de-constructed' rather than taken at face value. As we'll see later in the book, this has been especially evident in relation to 'race', gender and sexuality. To de-construct ideas of race, gender and sexuality is, in this context, to show that they rest on dubious concepts of biological determinism. Additionally, some human geographers argue that the specific social relationships and 'structures' in which individuals are enmeshed materially influence their physical and psychological being over the life-course. This means that a person's bodily and mental 'nature' cannot be understood in purely biological terms (e.g. as a product of their genes). Rather, it must also be understood as an effect of a person's positioning within wider social networks since these networks shape people both physically and psychologically.

This kind of human geography research is a very long way indeed from the investigations undertaken by environmental and physical geographers. Whether studying nature in the sense of the non-human world or as essence, human geographers have made a concerted effort to *de-naturalise those things conventionally seen as wholly or partly natural.* These geographers evaluate, as well as make, moral and aesthetic claims about nature (not just cognitive ones). And they often do so in a normative mode, passing judgement on that which they analyse. Some student readers will doubtless be surprised to read that several geographers study human nature – albeit in critical and non-naturalistic ways – as well as non-human nature. But this is not quite the departure from 'the geographical tradition' (Livingstone 1992) that it appears to be. As I will show in the next chapter, the 'experiment' inaugurated by the early geographers even extended to explaining human nature. This is often forgotten when potted histories of geography are written for students. Originally, geographers wanted to link the natural environment, human society *and* human nature together within one or other explanatory framework. Well over a century later we have come full circle, but with a twist. Today, as we'll discover later in this book, physical and environmental geographers refrain from talking about human nature altogether (preferring, instead, the focus on the physical environment). Meanwhile, a cohort of critical (or left-wing) human geographers want to talk about 'human nature' just as much as their forebears did. The crucial difference is that they want to do so in a de-naturalising way such that what we often call human nature is not as 'natural' as it appears to be (see Box 1.5).

Box 1.5 DE-NATURALISATION, DE-CONSTRUCTION AND DE-ESSENTIALISATION

In this book (especially Chapter 3) I will use a suite of terms that together imply a certain scepticism that natural phenomena are, in fact, natural at all. The most general of these terms is *de-naturalisation*. In the context of the chapters to follow, de-naturalisation means two things. First, it means recategorising that which seems to be or is claimed to be 'natural' and showing it to really be social, cultural and economic in character (or the result of social, cultural and economic practices). Second, it also means refusing to explain the characteristics of any given phenomena with reference to its supposedly 'natural' qualities (as when one might explain a person's intelligence with reference to their genes). *De-construction* is a more specialist term normally associated with a body of theorising called 'post-structuralism'. In this book I use it in a way that respects the spirit, if not necessarily the letter, of post-structuralist theorising. Here it refers to any attempt to reveal the 'symptomatic silences' that lie within any given claim about what nature is, how it works, what it does and how it should be treated. A symptomatic silence is the 'absent presence' of an idea, assumption or belief that helps to establish the meaning of a knowledge-claim yet without appearing to do so. For instance, in Chapter 3 we will see how environmentalists represent Clayoquot Sound in Canada as an untamed wilderness in need of protection and conservation. The problem with this apparently unproblematic claim is that local native peoples are depicted as being 'at one' with this wilderness – a view that reflects a romantic Euro-American belief in the 'edifying' power of pristine natural environments. To de-construct environmentalists' representations of Clayoquot is thus to show that their apparent stability and obviousness in fact rest upon culturally contingent distinctions between nature and culture, tradition and modernity, and the rural and the urban. Finally, *de-essentialisation* relates to the second of the three principal meanings of the term 'nature' discussed early in this chapter. A de-essentialising argument is a specific form of de-naturalisation. It questions the idea that any given phenomenon has a fixed and 'essential' character by virtue of its naturally given or determined properties.

To summarise, the geographical study of nature extends beyond the physical environment and involves human geographers as much as physical and environmental geographers. Despite initial appearances, geographers' research on nature could readily encompass most of the issues highlighted in the seven stories with which I began this chapter. This said, our discussion of geography and nature is not quite complete. Attentive readers will have observed the absence of any mention of the third definition of nature in this section. So let me now conclude this part of the chapter by quickly making amends. The notion of nature as an 'inherent force' may sound very abstract, even metaphysical. In other words, it may seem unlikely to be of interest to geographers. But appearances can be deceptive. In the first place, many physical geographers are interested in the inherent forces in the natural world that create everything from meandering rivers to glaciers. Far from being abstract and metaphysical, these forces include gravity, the conservation of energy and the increase of entropy, among others. They are forces mostly studied by the specialist sciences. Physical geographers have written whole books identifying and explaining the natural forces that are relevant to their research (e.g. Bradbury et al. 2002). In investigative and teaching terms, what usually interests physical geographers is the way these forces combine in specific real-world times and situations. For environmental geographers things are a little different. To simplify, an understanding of natural forces is more a part of their 'background knowledge'. It is relevant to their research but not something they would comprehend in the same detail as a physical geographer might do. As for human geographers, well things are different again. For obvious reasons, these geographers are not too interested in things like entropy or gravity (except, perhaps, as metaphors)! However, they *are* interested in the way some people *represent* nature as an 'inherent force'. For example, in 2001 Lynne Bezant – a 57 year-old British woman – became pregnant as a result of IVF. According to one of her critics, this entailed wrongly 'straying over nature's line' (Weale 2001: 3). Rather than take this criticism at face value, some human geographers would look not only at who said it, but how and why nature was invoked to make the criticism. They would ask what purposes it serves to *depict nature* as a force we ignore only at our peril (see Table 1.3 for a summary of this section).

Table 1.3 Geography and the study of nature

	Parts of geography	
Meanings of nature	Physical geography	Human geography
The non-human world	✓	✓ (Ideas about and alterations of the non-human world)
The essence of something	✓	✓ (Ideas and practices influencing bodily and mental 'natures')
An inherent force	✓	✓ (Ideas about nature as an inherent force)
	Environmental geography	

THE ROAD NOT TAKEN

We've already covered quite a lot of ground in this chapter. But we still have a little way to go before we fully understand the structure and aims of this book. It seems to me that two things distinguish geographers' contributions to our understanding of nature when compared to other academic disciplines. First, geographers investigate an unusually wide range of phenomena captured by the three meanings of the term 'nature'. Unlike relatively specialist subjects (say, chemistry), geography examines everything from the moral claims made by an organisation like PETA through to why lateral moraines are deposited by glaciers. Second, even though only environmental geographers actively try to combine them, geographers offer both social-science and physical-science perspectives on nature, as well as a humanities one. In other words, geographers employ more than one 'paradigm' (or framework of analysis) when investigating nature (Box 1.6). One of the key reasons for this is the above-mentioned breadth of things geographers study under the heading of 'nature'. For instance, it's clearly not appropriate to investigate urban heat islands in the same way as one might examine why the idea of 'wilderness' is so embedded in North Americans' imaginary.

This leads me to a third observation. When it comes to nature, there's a lot of mutual ignorance among geography's three main research and teaching communities. In other words, because they focus on such different

Box 1.6 PARADIGMS

The term 'paradigm' is famously associated with the historian of science Thomas Kuhn (1962). A paradigm is defined as 'the working assumptions, procedures and findings routinely accepted by a group of scholars, which together define a stable pattern of [research] . . . activity' (Johnston et al. 2000: 571). According to Kuhn, scholars in any academic discipline can be identified by the paradigm to which they subscribe. This implies that paradigms organise the way any given researcher investigates the world. This ranges from that researcher's philosophical beliefs (i.e. their assumptions about the nature of reality and how we can come to know that reality) to the specific laws, models and theories that they employ, to the specific investigative methods they favour, to the kinds of research questions they ask, to the kinds of real-world things they choose to study.

In geography, there was a fair amount of debate in the 1980s over whether and how the paradigm idea can help us to understand what geography researchers do. I don't want to revisit those debates here. I simply want to use the term 'paradigm' as a heuristic device to get student readers to recognise two things. The first is that paradigms are often what Kuhn called 'incommensurable'. In other words, paradigms are different 'worldviews' or 'languages' that cannot be readily translated into the terms of another. This means that, within any academic disciplines (like geography), one finds researchers investigating often the same aspects of the world but in radically different ways. Second, in principle, one or other paradigm can be dominant in a discipline at any moment in time. However, it's fair to say that no one paradigm is dominant in human geography. Though human geography is often described as a 'social science' this does not mean that one scientific approach dominates the field. Instead, one finds everything from Marxist to feminist to more scientific (or 'positivist') perspectives vying for dominance. Things are different in physical geography, where a broadly 'scientific' approach is accepted, even if it's far from homogeneous (see Chapter 4). Environmental geography is different again, because it often mixes and matches paradigmatic research practices from both sides of the discipline.

Sources: Kuhn (1962); Johnston et al. (2000); Johnston (2003: 12–25)

aspects of nature in such different ways, human geographers often have little detailed appreciation of what environmental and physical geographers do (and vice versa). I should say immediately that geography is not alone in this. Most other disciplines are comprised of several academic communities who know little about what their peers do. One of the aims of this book is to dispel some of the ignorance – especially among degree students – regarding the range of things studied by geographers under the rubric of 'nature' and regarding the different ways those things are understood.

How, then, have I sought to fulfil this aim? The obvious answer is that I've endeavoured, in the chapters that follow, to discuss all three parts of geography (environmental, human and physical). Though I'm a human geographer by training, Nature would have been a meagre book if it had considered only human geography alone. Equally, I do not limit my discussion to geographers' research on the environment (i.e. to only the first meaning of nature). Less obviously, a very literal approach to my topic would involve me detailing each and every way that geographers have investigated different aspects of nature. For instance, I would have to examine each sub-branch of physical geography, not to mention several branches of human geography. Since this would be infeasible (and make this book hopelessly long and indigestible), I've gone for a more parsimonious approach. My tack has been to identify the fundamental differences and commonalities in the ways geographers of all stripes investigate and understand nature. In other words, I do not spend a chapter explaining how, respectively, human geographers, physical geographers and environmental geographers approach the topic of nature (though I do, admittedly, discuss their contributions separately throughout the book). Instead, I identify broad similarities and differences within and between these three research and teaching communities. It is these convergences and divergences that organise the chapters that follow the next one.

This brings me to you, the reader. How should this book be read? I ask this question because the expectations of readers thoroughly condition how they digest a text. What are *your* expectations? If you're a student you may be looking for me to explain a few of the 'truths' about nature discovered by geography's three main research communities. Equally, if you're a professional geographer you may be looking for a gentle introduction to research findings in parts of the discipline outside your area of expertise. The assumption both types of readers might make is this: *Nature* will tell you what those different bits of nature that different geographers study are 'really like' according to current wisdom.

If you've made this assumption then I want to challenge it. If your expectations resemble those above then I want to question them also. Let me explain. In the second section of this chapter, 'Knowledges of nature', I made mention of the numerous organisations, institutions and professions that produce knowledges about nature (in all the meanings of that term). In the first section, 'Tales of nature', I illustrated the sheer pervasiveness of what we call nature in our collective discourse and practice. I now want to draw an important inference from this, as follows: *the power to say what nature is, how it works, and what to do (or not to do) with it is enormously consequential for people and the non-human world.* Those who possess this power can materially influence the lives of billions of people, not to mention the whole gamut of animate and inanimate phenomena that surround us. Fundamentally, it is the power to have *one's knowledge-claims taken seriously by significant parts of (or even most of) any given society.* This raises the question of why some knowledge-claims become widely accepted, while others barely get noticed.

ACTIVITY 1.3

Try to answer the question posed above. Think about a specific belief concerning nature that is now taken for granted. Ask yourself: what are the reasons this belief has become widely accepted?

I can think of at least two reasons why certain knowledge claims – not just about nature, but about *any* topic – are regarded as legitimate ones. First, in any society certain knowledge-producers are able to claim that their knowledge is (or aspires to be) especially 'truthful'. In most countries today, academics in general, and those calling themselves 'scientists' in particular, make this claim (even if they don't always do it loudly). The declaration that one's business is the production of truthful, accurate or otherwise objective knowledge is a powerful one that not all knowledge-producers can make. For example, while a tabloid newspaper may be very widely read, its readers are under no illusions that the knowledge disseminated by the paper is particularly accurate. Second, the ability to claim the mantle of truthfulness is often allied with the ability to instil trust in one's audience. Trust is, in essence, a social relationship. It entails one party believing that another – on which it relies for something – will say or do certain things according to certain standards. Trust is a very real but intangible thing. Those knowledge-producers that are trusted have an obvious edge over those that are not. For instance, in radio broadcasting, the BBC's World Service news bulletins are among the most trusted globally. This is because, more than many radio stations, the BBC has developed a reputation for fair and accurate reporting. The precise reasons for that are complicated. The point, simply, is that once a reputation like this has been established it can be used to great effect.

I talk about truth and trust because it would be all too predictable for some readers of this book to assume that geographers are in the truth-business and therefore to be trusted implicitly. Against this, I prefer to see geographers – and all academics – as *using the public's belief that they epitomise the two Ts to get the knowledge they produce taken seriously*. There is no better illustration of this than the process of teaching. One of the reasons why students learn the things their university teachers ask them to learn is because they have been taught to believe that their instructors are reliable experts. Instructors can use this belief to get students to imbibe – without dissent – certain bodies of information rather than others. In this sense, all teachers are 'gate-keepers' of knowledge. They use the authority that their claims to truthfulness and trustworthiness give them to *license* certain knowledge-claims and *censor* others.

In short, I believe that we should treat *all* the nature-knowledges that *all* three types of geographers produce as *representations of nature* and nothing else. In other words, I insist that we should not assume that academic disciplines offer us a privileged insight into nature's 'real workings', or the way societies interact with the environment, or the way other people's claims about nature are phrased and used. In geography's case, I think it's wrong to presume that physical and environmental geographers' research is about a 'real nature' and therefore objective and accurate. Equally, I think it's wrong to assume that when human geographers interrogate claims about, and practices upon, human and non-human nature, they offer neutral insights into their subject matter. Instead, I think it's more productive to regard all the nature-knowledges that geographers produce as *depictions* whose truthfulness is an open question. Rather than remain in thrall to the two Ts, we should ask the question: What gives certain knowledge-producers the ability to claim truthfulness and trustworthiness as part of their repertoire?

To answer this question in geography's case we'd have to undertake a sociological analysis of how the discipline has used its university status to gain the ear of generations of students and countless user-groups outside higher education. Obviously, I do not have space here for such an analysis. It's enough, I think, to lay down the following challenge to readers of this book. Rather than read *Nature* hoping to find out more about what nature *is* I'd invite you look for something else. I'd invite you to regard geography as one of several knowledge-producing domains that tries to convince you that its claims about nature are legitimate ones. This reinforces my earlier point that knowledges of nature are not the same as nature itself, even though they are always *about* those things classified as 'natural' phenomena. Geographers produce *understandings of nature*: knowledge, not the reality itself. It is an open question whether those understandings are true or false, good or bad, accurate or partial.

At this point some readers may feel nauseous, distressed or otherwise annoyed because of all my questioning of what geographers have to say about nature. So let me be clear. I am not doubting the rigour or honesty of geographers' inquiries into those things we describe as 'natural'. What I am saying is that the knowledge geographers produce is part of the process whereby certain actors get to decide how we, in the wider society, should understand nature and, even, what the term applies to. The question to ask about geographers' various claims about those various things we call 'nature' is not 'are those claims true or valid?'. Rather, what we need to ask is: What sorts of thoughts and actions are geographers' different knowledge-claims about nature designed to achieve? This very practical question gets us examining what's sometimes called the 'performativity' of all knowledge. Particular knowledge-claims always have consequences, especially when they're hegemonic ones. Like other academic disciplines, geography is in a privileged position to shape wider understandings of the natural world

– despite the BSE crisis and other doubts about the trustworthiness of what professional researchers say. So how have geographers represented nature and to what particular ends? Put differently, in Nature I want to examine the *ideas of nature* in which geographers have a considerable investment.

NATURE IS DEAD! LONG LIVE NATURE!

Once we distinguish ideas of nature from the things they refer to we can make an apparently startling claim: namely, that there is no such thing as nature! Nature is simply a name that is 'attached' to all sorts of different real-world phenomena. Those phenomena are not *nature as such* but, rather, *what we collectively choose to call 'nature'* (Urban and Rhoads 2003: 220). In this sense, nature does not exist at the ontological level (that is, at the level of material reality). If you think again about the seven stories with which we began this chapter, it's clear that a range of qualitatively *different* things are being encapsulated by the *same* label. Arguably, the only reason these various things seem to be similar is because they share a common name, not because they really have anything (or much) in common. The things we call nature undoubtedly exist. But it is entirely a matter of convention that we group them together under the one term. Even if the term isn't explicitly invoked to describe them, it is clear that it's nonetheless there in the background.

So when geographers talk about nature in their research and teaching (either explicitly or implicitly) we need to understand that they are not talking about nature but that which they *call* nature. If effect, nature is *made real* only because geographers – and many other actors in society – choose to talk about all sorts of things *as if* the word used to describe them *was* those things. What conclusions can we draw from this? One is that there is no such thing as 'the right word' to describe any real entity. Words are attached to things purely by convention. They cut into the connective tissue of the world and isolate out 'chunks' of it for our attention. Another conclusion we can draw is that names matter in the sense that the meanings of those names colour how we understand, and behave towards, the things they refer to. A recent high-profile 'scandal' offers a dramatic example of this in the realm of public affairs. In 2003 a well-known, male, British television presenter was implicitly accused of being a rapist by an equally well-known Swedish television presenter. Gossip, rumour and off-the-record briefings led to his name entering the public domain. Though he was never formally named nor proven guilty, the enormous semantic weight of the word 'rapist' was enough to bring his career grinding to halt. Regardless of his innocence or culpability, once this word became attached to his person it had material effects on him, his family and his relations with others. As Agnew et al. (1996: 8, emphasis added) rightly observe, words are not simply 'a medium for conveying meaning but the *producer*[s] of meaning'.

So what can we say about the word 'nature', a key term in geography and the subject of this book? The first thing to say is that, in its three main meanings, the word nature 'encourages us to ignore the context that defines it' (Cronon 1996: 35). The main meanings of the word 'nature' all divert our attention away from the fact that it is a word not reality itself. After all, each of these meanings refer to that which is supposedly given, unalterable or pre-existing. Second, like all words 'nature' is a *signifier* that possesses one or more *signifieds* that are, in turn, attached to all sorts of different *referents*. A signified is the meaning of a word (or sound or image). A referent is the particular real-world thing that the signified denotes.

signifier (word) → signified/s (meaning/s) → referent/s (real-world phenomena)

Third, unlike most concepts, nature is remarkably *polysemic*. In other words, it has multiple signifieds and countless referents – what cultural geographer Kay Anderson (2001: 71–2) calls 'a wildly elastic range of designations'. Nature is a portmanteau word or what social scientists call a 'chaotic concept'. The term's complexity derives precisely from the jumble of meanings and referents we've come to associate with it (Figure 1.4). John

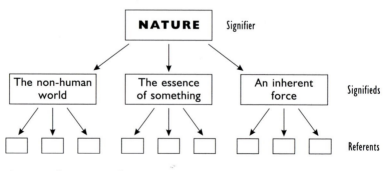

Figure 1.4 The concept of nature

Habgood (2002: 118) provides us with a useful analogy. In the sentences 'James had a fast car', 'James had a fast wife', 'James had a fast', one word means three quite distinct things and refers to very different phenomena. The word 'nature' is similarly promiscuous (Box 1.7). Strictly speaking, this means that geographers and others do not produce knowledge of nature in the singular but, rather, *natures* in the plural. Understanding the word is

---

Box 1.7   THE COMPLEXITIES OF A CONCEPT: NATURE

The following are a set of key quotes about the familiarity yet complexity of the term 'nature':

- 'It's complexity is concealed by the ease and regularity with which we put it to use in a wide variety of contexts. It is at once both very familiar and extremely elusive . . . an idea which most of us know, in some sense, to be so various and comprehensive in its usage as to defy our powers of definition' (Soper 1995: 1)
- 'The word nature is perhaps the most complex in the [English] language' (Williams 1983: 219)
- 'An immediate problem with the word "nature" is that it has multiple and overlapping meanings . . . Context can tell us a great deal about the shade of meaning intended' (Habgood 2002: 1).
- 'We cannot fall into the trap that this word has laid for us' (Cronon 1996: 36)
- 'Nature is a word which nowadays must be compulsively draped in scare-quotes' (Eagleton 2000: 83)
- 'The concept of nature has accumulated innumerable layers of meaning . . . Nature is material and it is spiritual, it is given and made, pure and undefiled; nature is order and it is disorder, sublime and secular, dominated and victorious; it is a totality and a series of parts, woman and object, organism and machine' (Smith 1984: 1)
- 'The idea of nature contains, though often unnoticed, an extra-ordinary amount of human history' (Williams 1980: 67)

manifestly not a case of identifying its 'proper meanings' and its 'proper referents'. This is something that the Welsh cultural critic Raymond Williams appreciated many years ago. As he put it:

> Some people, when they see a word, think that the first thing to do is to define it. Dictionaries are produced . . . and a proper meaning is attached. But while it may be possible to do this, more or less satis-factorily, with certain simple names of things, it is not only impossible but irrelevant in the case of more complicated ideas. What matters in them is not the proper meaning but the history and complexity of meanings . . .
>
> (Williams 1980: 67)

Following Williams, we can now see that all three main meanings (signi-fieds) of the term 'nature' are purely conventional not once and for all 'correct'. Likewise, there is nothing 'natural' about the fact that the term refers to all the particular things it does and not to others. If we want to know what nature is and why we value it in the ways we do we must look not to nature itself but to our ideas about nature. A telling example is provided by John Takacs (1996) in his book *The Idea of Biodiversity* (Box 1.8).

This discussion further confirms a point made in the previous section of the chapter. The claims that geographers make about nature are part of an ongoing process whereby the very meaning/s and referents of the term 'nature' are up for grabs. It's important to dispense with the idea that the term 'nature' innocently describes certain things that geographers then go out and study in detail. Rather, geographers' research and teaching is a part of the process where the meaning/s and referents of the term 'nature' become 'solidified' or 'fixed' at the societal or subsocietal level. Definitions of 'nature' do not precede the efforts of geographers and others to deter-mine what nature is and how, in practical, moral or aesthetic terms, to use it.

At this point I need to deal with a potential problem facing any author who wishes to analyse nature as one of geography's key concepts. The problem is this: many geographers prefer not to use the term 'nature' in their writing! For instance, many physical geographers favour the term 'environment' because, for them, the word 'nature' has quasi-romantic or mystical connotations of a 'higher power'. Likewise, many geographers are

Box 1.8 THE INVENTION OF BIODIVERSITY

The loss of biodiversity is currently a major environmental issue worldwide. It's estimated that humanity has discovered only a fraction of the naturally occurring species and habitats that exist on earth. At the same time, it's believed that many of these unknown species and habitats are being irretrievably destroyed by urbanisation, land clearance, agriculture, logging and road construction (to name but a few). What is biodiversity? According to conservation biologists, it describes the *number* and *variety* of plant, animal, insect and microbial species, as well as (at the subspecies scale) the number and variety of genetic traits and (at the supraspecies scale) the number and variety of habitats and ecosystems. Tropical countries, like Cameroon, are among the most biodiverse places on earth, while cold and dry countries are much less biodiverse. Countless books have been written about biodiversity, and biologists, agronomists, plant scientists, forest managers, geographers and environmental scientists are just a few of the professional researchers who take a keen interest in it. Biodiversity seems undeniably real and the current loss of biodiversity seems equally undeniable if one believes commentators like Norman Myers (author of the famous book *The Sinking Ark*, 1979). Indeed, if it weren't the case the United Nations would not, presumably, have coordinated efforts to create the global Convention on Biodiversity during the 1990s. Today this Convention has more than 100 signatory-countries from around the world translating its principles into national law.

Despite this, the American social scientist John Takacs (1996) has argued that biodiversity is an *invention*. Takacs's historical analysis shows three important things. First, he reminds us that the term 'biodiversity' did not enjoy common currency until relatively recently, entering the public domain only in the late 1980s. Second, he demonstrates that biodiversity has only become a well-recognised term because of the intensive efforts of a small number of conservation biologists with real concerns about the loss of genetic, species and habitat diversity. These biologists include Edward O. Wilson who, Takacs shows, used his fame and eminence to get a major publication on biodiversity commissioned during the 1980s (published as *Biodiversity* in 1988). Finally, Takacs shows that the term 'biodiversity' has brought together a set of what are considered to be 'natural things' within one unified conceptual frame that, previously, were looked at in rather different ways by both researchers and the wider public. In other words, while Takacs acknowledges that the natural world to which the term 'biodiversity' refers exists, he also insists that the term actively organises how that world is seen. In particular, he points to the normative dimensions of the term, whereby diversity is seen as 'good' and loss of/lack of diversity as 'bad'. He argues that these dimensions inhere not in biodiversity itself but, rather, reflect the values of biodiversity's champions. In this way, Takacs argues that people's values are surreptitiously passed off as values of nature. Arguably, 'biodiversity' has become a *hegemonic idea* in many scientific and policy circles (see Box 1.4).

Sources: Takacs (1996); Guyer and Richards (1996).

interested in specific phenomena – like precipitation or evaporation – that do not require formal use of the terms 'nature' or 'natural' to characterise them. So if I were to limit my analysis only to the work of those geographers who use the term 'nature' explicitly and formally then *Nature* would, in truth, be quite a slim volume. So how do I deal with this problem? And how do I justify discussing research where the term 'nature' does not enter the discourse? My 'solution' (that's what one can call it) is to do two things. First, I follow Kenneth Olwig's (1996) lead. In an essay on the concept of nature in geography, Olwig (1996: 87) shows that it is often 'a ghost that is rarely visible under its own name'. This seemingly enigmatic claim draws our attention to nature's numerous 'collateral concepts' (Earle et al. 1996: xvi). Collateral concepts are those whose meanings and referents overlap very closely with those of other concepts. Collateral concepts are mutually implicated and depend upon each other at some level for their meaning to be understood.

ACTIVITY 1.4

Can you list some of nature's collateral concepts? These concepts are ones that involve some or all of the meanings and referents of the idea of nature.

How many collateral concepts did you manage to identify? I've already mentioned one of nature's main sibling ideas (the environment – a concept which is also examined in this book series [see Endfield forthcoming]). Others include 'race', sex, biology, wilderness, countryside and rural to the extent that each of these is sometimes seen to have a 'natural' component (wholly or in part). For instance, the idea that humanity can be divided into discrete 'racial groups' frequently draws upon the idea of biological (i.e. natural in the second and possibly third senses of the term) difference. Once we appreciate that nature is a ghostly trace in several such collateral concepts we can expand the range of our analysis beyond those instances where nature is the stated object of discussion (Figure 1.5).

A second way we can justify expanding the range of our analysis is to recognise that it's just as important to examine what is claimed *not to be natural* as that which is. This may sound a little odd, so let me explain. Consider obesity. In recent years, newspapers have reported research that suggests a genetic cause for excessive weight gain. This can lead obese people and members of the general public to believe that obesity requires a medical cure (pills, injections, surgery or what have you). If, however, one argues that obesity is not natural – if one takes it out of this category – then

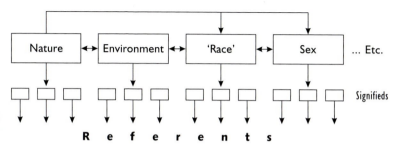

Figure 1.5 Nature and its collateral concepts

one's view of its causes and solutions alters drastically. One might look at a person's lifestyle or what kinds of foods match a person's budget given current prices. One might note that in Western societies obesity is pronounced among lower-income families. And one might conclude that obesity has social, cultural and economic causes such that medical solutions are misplaced or only partially valid. The relevance of this to geography is that, as we'll see later in the book, many geographers have sought to *re-categorise* those things that appear to be natural or to have natural causes. This attempt to establish the *non-natural* character of that which seems natural is an important part of determining where the boundaries between the natural and the social lie. Human geographers, in particular, have expanded our sense of where and why social processes, relations and structures are important. In many cases, they argue, nature is a *social construction* and thus not, in fact, natural at all (see Box 1.5 again).

SUMMARY

At this point it should be very clear what my approach to the nature geographers study is. I am not going to recommend one or other of understandings of nature that I discuss in this book – whether they come from human, physical or environmental geographers. Rather, I am going to treat these approaches as equally vigorous interventions in the important business of shaping wider understandings of, and practices upon, those things we call natural. I argue that we should evaluate these interventions less for their truth-value and more because of the kind of practical, moral or aesthetic projects they engender. For instance, I will ask what is to be gained in believing the claims of some human geographers that nature is a 'social construction'? Likewise, we will ponder what is lost if we treat the 'scientific' claims about nature made by many physical geographers with unbridled scepticism or else with uncritical enthusiasm. I hope it's now clear why I titled this introductory chapter 'Strange natures'. The way I've discussed nature in the preceding pages will have been triply strange for some readers. First, I have not discussed nature in the sense only of the environment (or non-human world). Second, this means that I intend, in this book, to discuss more than the nature investigated by physical and environmental geographers. Finally, the nature that geographers study (in all senses of the term) is not, I argue, to be confused with the things the term describes. Nature, I insist, is a concept or idea, not the real world of species, landforms and bodies. In keeping with the series of which this book is part, I treat nature as a 'key concept' in very literal and material terms. As we'll see in later chapters, other geographers see nature as an idea or 'discourse' too, but my tack is to take their claims about nature as *themselves* ideas about nature. The knowledge that geographers produce about those things we describe as natural is, I believe, both interesting and important. But student readers, in particular, must not treat this knowledge as a 'mirror' held up the natural world. In the next chapter I want to offer a brief historical survey of the changing 'nature' that geographers have studied and how this has altered 'the nature of geography'. Thereafter I want to explore in some detail the key – and often contradictory – under-standings of nature put forward by contemporary human, physical and environmental geographers.

EXERCISES

- Try to imagine not using the term 'nature' in your everyday conversation or in your degree studies. Do you think it would be possible to get by without the word or is it so ingrained in our language that it's indispensable?
- Take a few minutes to think about all the natural things that help to sustain your daily life. For instance, the next time you go shopping consider both the type and origin of all the natural things that, in either their original or processed form, go into your food basket.
- List ten things that you would describe as 'natural'. How diverse or similar are these things? Is the thing that makes them 'natural' the same in each case?
- Identify a film, an advertisement, a novel, a piece of art, a web site, a radio broadcast and a television programme where nature is a major topic. How, exactly, is nature portrayed in each case? Why do you think it is portrayed in these ways?
- Consider how describing something as 'natural' affects how you or others behave towards it. For instance, if you describe obesity as a 'genetic disease' then how would this influence your treatment of obesity if you were a medical doctor?
- Having read this chapter, go away and read Don Mitchell's (1995) essay 'There's no such thing as culture'. Like 'nature', the word 'culture' is complex. It has multiple meanings and myriad referents. Mitchell

argues that culture is not something real but, rather, an idea that has taken on a life of its own. See if you can apply Mitchell's analysis to nature. Ask yourself if nature is a similarly powerful *idea* – as I have argued in this chapter – or simply the real world of bodies, environments, species and the like.

FURTHER READING

There are several books and essays on the concept of nature going back to the historian R.G. Collingwood (1945). Two of the best recent ones are John Habgood's (2002) *The Concept of Nature* and Kate Soper's (1995) *What is Nature?* Habgood's book has a theological dimension but this does not detract from his main argument. Soper's book is denser than Habgood's and repays careful reading. Soper (1996) has written an essay summarising her book's thesis – this might be a good place to start for interested readers. Raymond Williams' (1980) essay on ideas of nature remains essential reading. The books by Glacken (1967) and Pepper (1984) are two of the few written by geographers about ideas of nature – but they mostly focus on ideas about the non-human world (the first definition of nature) and say less about the other meanings and referents of the term 'nature'.

For an insight into how ideas get invented by knowledge-producers and then gain a certain acceptance and influence see John Takacs's (1996) readable, fascinating book *The Idea of Biodiversity*. The first chapter of Tim Unwin's (1992) *The Place of Geography* argues that geography is a socially constructed discipline that generates its own distinctive bundle of knowledges about the world. Finally, for those readers who are challenged by my suggestions that 'there is no such thing as nature!', it's instructive to draw an analogy with Don Mitchell's approach to culture. 'Culture', like 'nature', is a complex word that describes a multitude of things. Mitchell (1995; 2000: ch. 3), subversively, argues that culture is not a real thing but, rather, a powerful idea that is strategically used by powerful groups in society *as* if it named things that were self-evidently 'cultural' in character (see the final Exercise above).

2

THE 'NATURE' OF GEOGRAPHY

'What is the relation between the nature of geography as a discipline and the nature that geographers believe ought to be the object of their study?'

(Olwig 1996: 63)

INTRODUCTION

In this chapter I want to offer a potted history of the ways in which geographers have understood nature since geography was founded as a university subject in the late nineteenth century. I want to do this for a couple of reasons. First, this chapter will set the scene for those that follow. It will introduce readers, in very simple terms, to some of the geographical knowledges of nature examined more closely later in the book. So, if you like, this chapter is a 'primer' for the rest of *Nature*. This explains why it's a rather long chapter (I recommend that it be read in three stages, taking a couple of sections at a time). Because I cover a lot of ground in this chapter I have chosen not to include many boxes or activities, since this would simply add to the length. Second, since the later chapters will focus on contemporary geographical understandings of nature, I wanted to ensure that readers could place this recent work within a longer temporal frame. *Nature* is not a book about the whole history of geographers' understandings of nature. Even so, it seems to me important to place present-day research within a long tradition of geographical inquiry into that which we call 'nature'. I should remind readers at the outset that this chapter – indeed this book – focuses on English-speaking geography. I have neither the space nor the expertise to discuss the history of university geography elsewhere in the world.

A discussion of how geographers have understood nature is, inevitably, a discussion about the 'nature' of geography. This is more than a wordplay. In this chapter we shall see that geographers' changing understandings of nature have been hard-wired to changes in the nature of their discipline. Paradoxically, this will lead us to appreciate that geography does not have (and has never had) a 'nature', if by that word we mean (in this context) a fixed character or identity (see Rogers 2005). I should warn readers that they will have to proceed slowly and carefully if they're to properly grasp the claims I make in this chapter. When I use the word 'nature' in the pages ahead it is always contextually. Depending on what's being discussed, the nature in question will be one or more of the term's principal meanings. In addition, readers will need to be alert to when I'm discussing nature in the cognitive, moral or aesthetic sense, and also when I'm talking about it descriptively or normatively. This is a lot to ask, I know! It's likely to induce the semantic equivalent of seasickness. But my excuse is that geographers have studied 'nature' in so many and varied ways that the one word inevitably changes its meanings depending on whose work I discuss. As I argued in Chapter 1, nature appears in geographers' work not only in its own name but also in the form of numerous collateral concepts. Henceforth in this chapter, I shall refrain from draping the word nature in scare quotes and simply leave it to reader to decipher its specific meaning in any given sentence.

Before I begin my potted history, I should confess that it is inevitably a partial and biased one. It is also too 'tidy', ignoring numerous loose ends and a good deal of complexity. In one sense this is the inevitable result of me having to cover human and physical geography, as well as the marchlands between them, in just one chapter. But even if I could write at greater length, Livingstone (1992: 5) is surely correct that *all* intellectual historians 'stage-manage' the facts of a discipline's past so that they never simply speak for themselves. Like Livingstone, I prefer 'contextual histories'. Such histories refuse to see academic disciplines as domains of pure rationality isolated from wider circumstances. As opposed to 'internalist' histories – which explain intellectual change only with reference to debates within a discipline – contextual histories look at how (i) wider currents of thought and (ii) wider social, cultural, economic and political circumstances together shape academic subjects over time and space. In short, a contextual approach explores the reciprocal relationship between disciplines and the wider context in which they are embedded.

By situating the changing ways geographers have studied nature within successive contexts we will see how the 'nature of geography' has rarely been stable. My main thesis is twofold. First, I'll argue that geography has always had a 'problem' with nature. Disagreements over what is and is not natural, and how best to study nature, have, I argue, been flash-points for the successive reconstitution of geography as an academic enterprise since the discipline's inception. Second, I want to suggest that contemporary geography has come full circle vis-à-vis nature from where it started more than a century ago – but with a twist. Victorian and Edwardian geographers sought to bring the study of the environment and human nature within one intellectual framework. In a sense, many of the early geographers were interested in 'outer' (non-human) and 'inner' (human) nature at some level. After a long post-1945 era where the study of nature was squeezed out of human geography, we are once again in a period where it is 'on the agenda' for an awful lot of geographers. Here's the twist though (to be explained in more detail later): in the main, human geographers take an emphatically *de-naturalising* approach to those things that are often thought to be natural – be they non-human species or even the human mind and body. Meanwhile, environmental and physical geographers are careful to limit their research to nature as environment and do so, in the main, in order to disclose its 'real character'. Consequently, we find contemporary geography 'divided' over its understandings of nature, with no prospect of a new grand theory to match that of geography's late-nineteenth-century founders. This, I conclude, is a good thing. It is neither practical nor desirable to comprehend those things we call natural within one or other overarching theoretical, methodological or evaluative framework.

BEGINNINGS

Historians of geography often trace the subject back to the likes of Herodotus and Ptolemy – an attempt, no doubt, to impress upon readers how ancient, and therefore *important*, geography is. My own story starts in the late nineteenth and early twentieth centuries, the period when anglo-phone geography was instituted as a university (and school) discipline. During this period three related things happened: the first professional geographers were appointed (like Halford Mackinder, Reader in Geography at Oxford University from 1887); the first geography departments were created (like that at the University of California, Berkeley, in 1898); and the first geography degrees were awarded to undergraduates and postgraduates (like the PhD granted by the University of Chicago in 1903). Establishing geography as a university subject was a slow, precarious business. However ancient the subject's lineage might be claimed to be, it required persistence and proselytisers to persuade universities to create a formal place for it within the academic division of labour. During geography's early years, a few key people – those with energy and vision – were able to shape the identity of this nascent discipline. That so few individuals could be so influential was a function of the simple fact that geography was a very small subject at this time.

So who were these individuals? How did they define geography? And why did they need to persuade universities that geography could be a 'proper' discipline, entitled to the same status as established subjects? Following Livingstone (1992), I want to focus briefly on three early geographers and situate their visions of the subject within a wider intellectual and socio-economic context. As we'll see momentarily, what they had in common was the belief that geography was an *integrative* subject. In Chapter 1, I discussed geography in terms of its three constituent communities. In the late nineteenth and early twentieth centuries, though, it's fair to say that geography was conceived as a holistic or synthetic university discipline. In other words, its early advocates did not see any sharp divide or difference between human and physical geography. Consequently, the label 'environmental geography' was not in currency since there was no perceived need for a 'middle ground' that would fill the gap vacated by the subject's two 'halves'. As Livingstone (1992: 173) puts it, 'This theme of connectedness, of the hanging-togetherness-of-things[:] . . . if geography had an independent disciplinary identity it was to be found here, in its capacity to integrate the disparate elements of world and life into a coherent whole'. As evidence of this, even some of the early publications that used the terms 'human' and 'physical' geography did not do so in any absolute or exclusive way. For instance, Mary Somerville's pioneering book *Physical Geography* (1848) belied its title by discussing human–environment relations within its pages. A generation later Halford Mackinder, William Morris Davis and Andrew Herbertson would each offer their own integrative visions of geography in order to get the subject recognised at university level.

Halford Mackinder (1861–1947)

In January 1887 the 26-year-old polymath Halford Mackinder gave a germinal lecture to the Royal Geographical Society (RGS) in London. Entitled 'On the scope and methods of geography', Mackinder's programmatic address sought not only to define geography but also to establish its importance as a field of research and teaching. For Mackinder (1887: 145), geography's role was to fill the 'gap' 'between the natural sciences and the study of humanity'. It was to be a 'bridging' discipline that studied human–environment relations. The immediate context for Mackinder's championing of geography was twofold. First, throughout the nineteenth century both the British public and successive British governments had a thirst for knowledge about other peoples, other places and other environments. The RGS actively supported numerous expeditions to Africa, the Antipodes and elsewhere – expeditions that were closely linked to the expansion of the British Empire. These journeys into 'unknown lands' employed topographic description, social survey, resource inventory, mapping, field sketching and comparative observation to generate a mass of factual information about non-Western peoples and environments. Important as this information was in the cause of colonial expansion, Mackinder rightly recognised that 'geography can never be a discipline' if it consisted merely of 'data to be committed to memory' (1887: 143). Second, at the time of Mackinder's lecture new disciplines whose hallmark was specialisation were gaining a foothold in universities. This was a problem for geography because its subject matter was so broad. Other disciplines could rightly claim that they already studied what geographers did or could do so in a more focused way. As David Stoddart (1986: 69) noted, 'Geography . . . appeared vague and diffuse, part belonging to history, part to commerce, part to geology'.

Together, the empiricism and sheer breadth of geographical knowledge meant two things for Mackinder. First, he needed to show that geography could 'trace causal relations' (1887: 145) and thus focus on *explanation* not simply description. Second, since he could not define geography in terms of its subject matter he had to emphasise its *distinctive perspective* on the world. This explains Mackinder's vision of geography as a 'unifying' subject that traces the interactions not only among human and physical phenomena but also between them. Geography was to make a virtue out of *not* being a specialist subject at a time when academic specialisation was the trend. If a wedge was being driven between the study of society and the non-human world during the late nineteenth century, geography would fill the nascent gap.

William Morris Davis (1850–1934)

Davis was a Harvard University geologist who, from the 1890s, had a major impact on American geography. Unlike Mackinder, his influence began not with a keynote speech but with a steady stream of research publications, textbooks, and memberships of university committees – as well as a role in the new Association of American Geographers (AAG). However, like Mackinder, he saw geography as that discipline which studied the 'relations between physiographic controls and ontographic responses' (Davis 1906: 70). As a geologist he was keen that the physical dimensions of geography not be assimilated into the natural sciences. His strategy was to emphasise 'the unbroken chain of causation linking the physical phenomena of the earth's surface, the organic realm and human society' (Leighly 1955: 312).

In light of this, it is ironic that Davis is best known as a pioneer of physical geography – and specifically geomorphology. His lifelong interest in the evolution of landforms was no doubt a legacy of his geological training. It was also a reflection of his American upbringing. Unlike Britain with its empire, the American experience of 'new territories' in the nineteenth century was dominated by westward expansion into the great plains and the Rockies. Though peopled by indigenous societies, the American west was a vast terrain of mostly natural landcapes. Describing and cataloguing those landscapes was a prime task of the United States Geological Survey. It was in this context that Davis, with his geological education, agitated on behalf of geography. Davis was well aware of the works of James Hutton and Charles Lyell. In the early to mid-nineteenth century, these had revolutionised understandings of the earth. They showed that the earth's surface and its underlying geology were the product of natural processes operating over very long timescales. Hutton's and Lyell's writings

challenged then-dominant views that God had created the earth just a few
thousand years ago and also questioned the belief that there was a 'divine
purpose' undergirding the natural world. Inspired by this non-theological
interpretation of earth history, Davis saw a niche for the study of present-
day physical environments. His own particular obsession was with visible
landforms (especially those of the eastern USA).Through field observation,
Davis pioneered 'denudation chronology' which envisaged initial geological
uplift, the action of fluvial erosion over time (process) upon the underlying
geology (structure), producing a physical landscape describable in terms
of its stage of development (youth, maturity, or old age). For Davis, the
end point of a 'normal cycle of erosion' was a fluvially eroded base-level
(a peneplain). He went on to modify his theory and extended it to marine,
arid and other environments.

Davis's influence on American geography was profound and yet strangely
double-edged. On the one side, his writings as a geomorphologist gave
an impetus to the development of physical geography as a distinct field of
study. On the other side, this tendency to sever the study of environment
from the study of society was counteracted in his teaching. Some of Davis's
students at Harvard went on to aggressively stress the study of human–
environment relations as geography's *raison d'être*. As we'll see below, they
did in deed what Davis only did in word.

Andrew John Herbertson (1865–1915)

Herbertson, like Mackinder, had a broad intellectual training in the sciences
of nature and of 'man'. He was recruited to Oxford by Mackinder in 1899,
the year his influential book *Man and His Work* was published. Despite its
title, this was not written in the George Perkins Marsh mould. Rather,
it stressed the influence of the physical environment on human societies
(not vice versa). Like Mackinder, Herbertson regarded an understanding
of the non-human world as an essential component in comprehending
how societies evolve. However, unlike Mackinder, Herberton accented
geography's role as the study of *regions*. In his agenda-setting essay 'The major
natural regions' (1905), Herbertson viewed the world as a physiographic
patchwork, each piece of which possessed a 'unity of configuration, climate
and vegetation' (1905: 309). For him, geography's role was not to explore
human–environment relations in general but, rather, their unfolding in
specific regional complexes. As one of Herbertson's famous admirers later

noted, 'it would be difficult to cite any other single communication which has had such far-reaching effects on the development of our subject' (Stamp 1957: 201).

Herbertson's vision of geography as the study of regions and of regional differences was consistent with the colonial origins of the discipline in Britain. What the RGS-sponsored expeditions of the nineteenth century had shown was that the world was immensely differentiated in both human and physical terms. Closer to home, the transition from rural-agricultural to urban-industrial society in Western Europe was overlaying a new pattern of geographical difference upon an older, long-established one. Like Mackinder, Herbertson realised that if geography was to win a place at the academic table it needed an identity that was both distinctive and respectable. The study of 'areas in their total composition or complexity' (Holt-Jensen 1999: 5) seemed to offer such an identity. No other discipline (except perhaps anthropology) could claim to study how the phenomena analysed separately by the 'systematic disciplines' combined in time and space. For Herbertson and those inspired by his vision, geography has unique among the disciplines in its focus on regional variation and its causes.

The evolutionary impulse

Mackinder, Davis and Herbertson – and a few other individuals – effectively 'invented' geography as a university subject in the anglophone world. They gave the discipline an identity and, with the help of organisations like the RGS, began the slow process of establishing it within the academy. With Mackinder we see the inauguration of what we today call 'environmental geography'. With Davis, we see the beginnings of physical geography, despite his Mackinder-like insistence that geography was a bridging subject. Finally, with Herbertson we see the beginnings of regional geography, something that is far less prevalent today than it was a century ago.

Leaving aside Davis's specialist interest in geomorphology, all three authors saw geography in holistic terms. But what kind of holism was this? The question is an important one because if geography was to be the causal discipline Mackinder wanted it to be then a substantive theory of how physical and human phenomena interacted was required. According to Livingstone (1992), that theory was furnished by a mixture of evolutionary biology, social Darwinism and neo-Lamarkianism. Let me briefly take each in turn. According to one commentator, the publication of Darwin's *The Origin of Species* 'caused a greater upheaval in [people's] . . . thinking than any other scientific advance since . . . the Renaissance' (Mayr 1972: 987). Darwin (1809–82) wanted to explain the multitudinous variety of life on earth. Was it a divine invention, as natural theologians believed, or the product of something else? And were the differences within and between species permanent ones? Darwin's answer to these questions had four main sources of inspiration. First, his voyage aboard the *Beagle* in the early 1830s had led him to observe the striking differences, as much as the similarities, among plant, animal and insect species in different parts of the world. Second, his interest in pigeons led him to observe that breeders were able to alter the physical characteristics of their birds over time through selective reproduction. Third, Darwin was much influenced by the famous essay of Thomas Malthus (1798) on 'the principle of population'. Malthus famously argued that human populations always expand up to and beyond the resource base available to them; for him this was a 'natural law'. Finally, Darwin's reading of Hutton and Lyell persuaded him that the history of lifeforms might be as long as that of landforms.

From his expeditionary observations and his studies of pigeon-breeding Darwin derived the metaphor of 'natural selection'; from Malthus's work he derived the idea of the 'survival of the fittest'; and from Hutton and Lyell he derived the idea of 'deep time' – the idea that change occurs slowly over the *longue durée*. Darwin's theory of biological evolution saw species as interacting with each and their wider environment. Over time, those species best able to adapt to their conditions of existence are most likely to survive and to produce offspring similarly well adapted. What Darwin showed was that one need look no further than the interactions between organic forms to explain species diversity. He also showed that, given enough time, biology was plastic not permanent. In short, *The Origin of Species* – a book published in numerous editions during the Victorian era – demonstrated that there was an order and direction at work in the natural world that could be inferred from empirical observation of species' characteristics. However, contra natural theology, this order and direction was, in Darwin's estimation, 'blind': it was the unplanned outcome of continuous processes of competition and adaptation.

Darwin's book was about the non-human world, even though it was not lost on him that people are biological, as much as social, animals. 'Social Darwinism' was a set of diffuse yet very real beliefs that gained currency in Victorian Britain and beyond. They maintained that whole societies (and classes or groups within them) were akin to species. In Britain, early sociologists and statisticians (like Herbert Spencer and Francis Galton) popularised the idea that competition within and among societies is 'natural' – an idea that justified European colonialism as much as a belief that in any society the 'fittest' (mentally and physically) rise to the top of the hierarchy. Finally, in the late nineteenth century, several of Darwin's acolytes (and some of his critics) resurrected the evolutionary doctrines of Frenchman Jean-Baptiste Lamarck. Neo-Lamarckians maintained that evolution proceeded more quickly than Darwin had allowed and was not subject to Darwin's 'random variation'. They insisted that the qualities acquired by an organism during its life-experience could be directly transmitted to its progeny and that will, habit or environment drove evolution forward, not a 'blind' process of competition, variation and adaptation.

What has this *mélange* of evolutionary thinking got to do with late-nineteenth- and early twentieth-century geography? A good deal as it turns out. According to Livingstone (1992) evolutionary theory – domesticated to their own needs – furnished the early geographers with a means of bringing humans and the environment within a single explanatory framework. In its 'strongest' version this framework proposed to link non-human nature with human nature (bodily and mental) and human society. This was was most obvious in the works of Davis's students, such as Ellen Semple and Ellsworth Huntington, where the neo-Lamarckian strain was strong. Semple's (1911) tellingly titled *Influences of Geographic Environment* and Huntington's (1924) *The Character of Races* made causal links between physical environment, the bodily and mental capabilities of different human groups (Caucasians, Blacks, Aborigines etc.) and the levels of 'civilisation' of these groups. Even as late as 1931, it was not unusual to encounter the following statements: 'Psychologically, each climate tends to have its own mentality, innate in its inhabitants and grafted onto its immigrants' (Miller 1931: 2).

This sort of 'environmental determinism', as it became known, was a curious doctrine. It awarded the physical environment the status of an independent variable, while making human nature and human society dependent variables. Yet, for all its determinism, it did allow room for change and manoeuvre in human biology and ways of living. As Semple (1911: 2) acknowledged, humans were seen as 'shifting, plastic, progressive

[and] retrogressive' precisely because of the neo-Lamarckian belief that biological traits and social improvements could quickly be passed from generation to generation. In practical terms, this belief found wider expression in the eugenics movement (eugenics means the science of 'racial improvement'). The popularity of eugenic thinking and practice in Europe and North America in the early twentieth century no doubt gave succour to the likes of Semple (Box 2.1). Aside from its contradictions, environmental determinism was blatantly racist and imperialist at times. It ranked societies worldwide on a scale of development that equated 'uncivilised' societies with 'harsh environments', and it implied (illogically) that most Caucasians had somehow escaped the determining force of their own environments in a virtuous cycle of biological and social evolution.

Box 2.1 THE EUGENICS MOVEMENT

Narrowly conceived, eugenics is the theory and practice of human biological 'improvement'. More broadly, it is the belief that there are certain 'imperfections', 'problems' or 'flaws' – mental and physical – that should be engineered out of existence for the greater good of society. Such 'engineering', these days, can in principle be achieved at the genetic level. In the past, though, it was more likely to be achieved by controlling which individuals and groups could procreate (through sterilisation programmes for instance). Eugenics became popular in the early twentieth century in both Western Europe and North America. It arose in the wider context of social Darwinism and was coincident with the discovery of 'rules of heredity' in both plant and animal breeding and hybridisation. At its worst, eugenics was racist, ethnically elitist and socially exclusionary. For instance, by the 1920s, many Western governments had introduced sterilisation policies in order to prevent people suffering physical deformities, mental retardation, epilepsy, schizophrenia and other chronic illnesses from having children. The idea was that a national population could 'iron out' these biological flaws over time. In Nazi Germany eugenics became linked with the extermination of Jews, homosexuals and other supposedly 'inferior' groups of people. This demonstrates how value-laden eugenic beliefs inevitably are. For in order to determine what *counts* as an 'undesirable' mental or physical condition someone has to make a decision about where the line between 'normal' and 'abnormal' 'human nature' lies – and that decision will reflect the values and beliefs of the decision-makers. In other words, we cannot simply 'read off' a person's worth, happiness or well-being from the 'facts' of their mental and physical inheritance. Though the word 'eugenics' is rarely used today, some worry that the new-found power of biomedicine to intervene in the workings of the human body is reviving eugenics 'by the back door' (Duster 1990).

To be fair, not all the early geographers can be accused of being environmental determinists. Herbertson, for example, was certainly influenced by Lamarck through his teacher Patrick Geddes. But his interest in the regional specifics of human–environment relations moderated after *Man and His Work*, where he more or less 'read-off' modes of life (pastoral, nomadic, industrial etc.) and 'regional character' from the natural-resource endowment of an area. Like another of Geddes's students, Herbert John Fleure (1877–1969), Herbertson became a 'possibilist' over time – in the vein of the French school of regional geography centred on Paul Vidal de la Blache. Possibilists, as the name suggests, believed that the physical environment offered opportunities to, as much as it placed constraints upon, its human inhabitants. This granted human society, if not necessarily human biology, a degree of relative autonomy. It also meant, among other things, that possibilists were less guilty of racist and supremacist prejudices than their determinist contemporaries. Fleure, for example, celebrated the physical and human diversity of regions rather than simply arraying them on a single scale from 'harsh and barbaric' to 'temperate and civilised'. Fleure's regional geography did not provide ammunition to those seeking to justify the conquest or exploitation of some regions on the grounds of the supposed 'inferiority' of their peoples.

Reflections

Clearly, the question of nature was absolutely central to the founding of geography as an academic discipline. Indeed, it's fair to say that geography was, in its origins, very much a *naturalistic subject*. By this I mean two things: first, that 'natural' phenomena were principal objects of analysis and, second, that these objects were used in explanations of the 'non-natural' dimensions of life (like culture). So not only was nature a prime focus of geographical research, it was also granted prime causal importance in explaining various other geographic phenomena. There were, if you like, 'two natures' central to geography at this time: a 'first nature' (the environment) which imposed itself upon 'human nature' (body and mind) more or less restrictively depending on the humans in question. In turn, both elements of nature were seen to have a causal influence on how society, culture and economy were organised among different human groups.

Yet, for all its centrality, it seems to me that nature was also a double-edged sword for the early geographers. On the one hand, these geographers used the topic of nature to establish the distinctiveness of their perspective on the world. They were not, to be sure, the only ones studying nature (non-human and human). And so their way of marking out academic territory was to stress their holistic and integrative focus on the particular *relationships* between particular natural environments and particular forms of human nature (physical and psychological) and society. On the other side, though, this holistic and integrative perspective was arguably at the heart of geography's intellectual weaknesses. In the first place, it meant that the geographers' research agenda was overwhelmingly large. To make good on the Mackinder–Davis–Herbertson vision, geographers would not only have to know an awful lot about everything from soils and vegetation to industry and culture. They would also need to understand how all these things interacted in a causal sense. Second, it rapidly became clear that both human–environment and regional geographers preferred description, metaphor and speculation over well-justified explanation in their written works. The sheer scale of their intellectual projects meant that they were generally unable to specify the exact causal connections between people and environment. Their work was, often, more in the impressionist mode of humanities subjects like English and art history. Geographers were thus arguably sandwiched between a rock and a hard place. As the specialist disciplines appropriated more and more intellectual territory, geography was left in the invidious position of studying *everything* that the other subjects combined did. In either the human–environment or regional modes, this was always going to be difficult.

Third, this left the option of making geography the study of the natural environment alone (or at least separately from society). As we've seen, this was the option that Davis pursued, despite himself. It had also been pursued before him by Somerville and also by Thomas Huxley (1877), whose book *Physiography* examined the connections between the biosphere, hydrosphere, lithosphere and atmosphere. This vision of geography as an earth science, because of Davis's great influence, actually became a live one – especially in early twentieth-century American geography. But it was always risky because a nascent discipline (or subdiscipline) of 'physical geography' was inevitably going to face competition (and even hostility) from geologists, plant biologists, zoologists and other specialists interested in the non-human world. Yet, despite these various problems, geography slowly – but nonetheless surely – prospered as a discipline in Britain, the USA, Canada, Australia and New Zealand as the nineteenth century gave way to the twentieth. As we now turn to the second main part of our history, we'll see that the 'problem of nature' present at the moment of geography's founding did not go away. This problem, as I'll explain, was instrumental in greatly altering the 'nature of geography' by the mid-twentieth century.

EARLY TWENTIETH-CENTURY DEVELOPMENTS

'To occupy or vacate the "middle ground"? That is the question'

By the 1920s, some of the tensions surrounding whether and how to incorporate nature into geography's *raison d'être* were bubbling to the surface. Those clinging to a holistic conception of geography were doubtless dismayed by the desire of some to develop a 'systematic' or 'general' geography. Inspired by Davis, many geographers who had a natural-science training eschewed the broad synthetic ambitions of the regional and human–environment approaches. Instead, they sought to become environmental specialists. Because of Davis's prodigious output, geomorphology (the study of landforms) fast became the major branch of physical geography in the anglophone world (and arguably remains so to this day). However, biogeography (with soils) and climatology (with meteorology) also gained an early impetus. In the first case, the ideas of Fredric Clements (1916) pioneered the classification of vegetative communities and their analysis in terms of succession, climax and dynamic equilibrium with the surrounding environment. Meanwhile, the American C.F. Marbut (1935) translated the path-breaking work of Russian soil taxonomists and analysts for English-speaking geographers. Marbut then went on to publish his own work on US soils. In the second case, the research of Norwegian scientists like J. Bjerkness in the 1920s showed that it was possible to classify and describe the life history of distinct air masses and weather systems.

One of the obvious gains of this sort of specialised geography was that it held out the prospect of physical geographers becoming environmental experts. It not only circumvented the 'jack-of-all-trades' problem associated with the regional and human–environment conceptions of geography. It also promised to offer fairly *precise* descriptions and explanations rather than the often vague, woolly, impressionistic analyses offered by the likes of Semple. However, there were two obstacles to the development of physical geography at this time. First, many of the early geomorphologists, biogeographers and climate geographers worked in non-geography departments, in part because there were still few geography departments in existence. Second, the risk of vacating the 'middle ground' occupied by regional and human–environment geographers was that geography became vulnerable to assimilation by cognate disciplines. For all the flaws of the evolutionary approach to nature–society relations, a study of these relations allowed geography to claim a distinctive place within academia. But once Davis and others began to 'ruptur[e] . . . the newly stitched sutures' (Livingstone 1992: 210) holding geography together, there was the risk that physical and a yet-to-be-created human geography could not survive on their own.

Strangely, early twentieth-century geography failed to champion a theme that was both topical and that might have offered a preferable way of examining human–environment relations when compared with evolutionism in general and environmental determinism in particular. It was the theme of what a later book by geographers, in its title, called 'man's role in changing the face of the earth' (Thomas 1956). This theme had already been broached by George Perkins Marsh in his book of 1864. In the United States, it was a theme at the heart of the new 'conservationist' and 'preservationist' movements associated with John Muir and Gifford Pinchot that emerged around the turn of the twentieth century. Here, as elsewhere in the Western world, the environmental impacts of rapid population growth, mass industry, commercial agriculture and urbanisation were starting to become apparent. The first modern cases of resource over-exploitation came to light, and it became clear that in some societies the causal arrows linking people and environment ran from the former to the latter, not vice versa. Quite why geography in this period did not capitalise on the 'human impact' theme is hard to explain. Aside from a few isolated studies – such as Jacks' and Whyte's (1939) *Rape of the Earth* and Cumberland's (1947) research into soil erosion – it was barely evident in pre-1950s geography. Perhaps the intellectual impact of evolutionary thinking was so large that there was simply no space for ideas about humanity's domination of the physical environment.

Even so, a number of geographers chose to emphasise geography's 'bridging' function well into the 1930s. But as the intellectual weaknesses, not to mention the sometimes crude moralising, of environmental determinism became evident, these geographers looked for other ways of thinking about human–environment relationships. In Britain, where environmental determinism was never as strong as it had been in the USA, regional geographers produced a steady stream of monographs that were rich on detail about physical and human landscapes but which avoided grand pronouncements about the causal connections between the two. Among these geographers were Darryl Foord, Percy Roxby and H.C. Darby. In the USA, the President of the AAG – Harlan Barrows (1923) – talked about geography as the study of 'human ecology' (a term coined by Chicago geographer J.P. Goode in 1907). Influenced by Clements's ecology and the ideas of Ernst Haeckel (with whom the term 'ecology' is originally associated), Barrows saw different societies as adjusting, adapting and modifying themselves in relation to their environmental conditions. This was not environmental determinism but, rather, an open-minded commitment to studying the two-way (dialectical) connections between a conditioning physical world and responsive human societies.

The human side of the dialectic that interested Barrows was the concern of an American geographer arguably as influential in his own country's discipline as W.M. Davis had been earlier in the century: namely, Carl Sauer. Sauer, a Berkeley geographer, was greatly influenced by the antideterminism of anthropologists Franz Boas and Alfred Kroeber. For him physical geography played 'an important role in providing the background to human activities' (Unwin 1992: 97). But he resisted the idea that these activities could be directly explained in terms of the physical environment. Like Boas, Sauer was influenced by the seminal work of the European philosopher Wilhelm Dilthey. Dilthey, departing from the extension of Darwin's ideas to the human realm, distinguished between the 'sciences of nature' and the 'sciences of man'. The former, he argued, sought *explanations* of a more or less orderly natural world. The latter, by contrast, took human culture and thinking as their subject matter and thus quested more for *understanding*. In effect, Dilthey drove an ontological wedge between people and environment: because they were *different orders of reality* they needed, Dilthey maintained, to be studied in different ways. Building on this, and on Boas's and Kroeber's commitment to fieldwork over sweeping theoretical generalisations, Sauer published his influential essay 'The morphology of landscape' in 1925. In it he defined geography's subject matter not as human–environment relationships but, rather, as the visible consequences of people's actions upon the landscape. His kind of cultural ecology or cultural landscape study took it as axiomatic that people routinely 'transform . . . the natural landscape into a cultural landscape' (Sauer cited in Livingstone 1992: 297). But unlike Marsh, Sauer was not much interested in the destructive, epochal transformations wrought by industrial societies – at least not until later in his career. Rather, his empirically driven, rather atheoretical agenda focused on rural and historical landscapes in all of their cultural particularity. This no doubt reflected the influence of Boas and Kroeber, who were fascinated with non-Western peoples past and present. In the work of Sauer and his many students, geography was presented as a 'chorological discipline' that examined 'culture areas' in which different natural landscapes were slowly transformed by different peoples in different ways. Morally, Sauer's landscape morphology possessed none of the absolutism of the environmental determinists and exhibited the generosity – the celebration of geographical difference and particularity – found in the regionalism of Geddes, Fleure and the later Herbertson.

If Sauer's work inclined more to the human side of the people–environment relationship, this did not mean that 'human geography' was yet a recognised part of the discipline. To be sure, some regional geographers studied the human dimensions of a territory more than its physical ones (e.g. Fleure 1919). Meanwhile, others studied single human aspects of regions, and this is how both economic (or 'commercial') and political geography came into being. Mackinder, for instance, was fascinated with how the world's physical geography influenced inter-state relations – both diplomatic and military. Mackinder saw the irregular geography of land, sea and resources as the flashpoint for inter-state struggles over survival and prosperity (see Mackinder 1902). In a related vein, the Briton George Chisholm and the American J. Russell Smith also looked at how the uneven quality and quantity of natural-resource endowments across space influenced the geography of economic activity. Chisholm's multi-edition *Handbook of Commercial Geography* (first published in 1889) and Smith's *Industrial and Commercial Geography* (1913) both showed how different societies had built different industries on a variable resource base. Yet, despite these efforts, as late as 1948 Isaiah Bowman – a leading light of his disciplinary generation and remembered today for his writings on geopolitics – told Smith that human geography could never be an independent arm of geography (Smith 1987: 162). Sauer's influential research seemed only to confirm this because it insisted on linking the study of cultural attitudes and practices to quotidian modifications of natural landscapes.

Nature in geography and the nature of geography

Let me summarise. On the eve of the Second World War, anglophone geography comprised four main strands of research and teaching, the first and last of which were overshadowed by the other two. There was a nascent physical geography, dominated by geomorphology, which was more evident in the USA than elsewhere. Second, there was regional geography, increasingly distancing itself from environmental determinism, especially in Britain. Third, there was a continuing tradition of human–environment geography, also fast ridding itself of determinist baggage, and increasingly subsumed within the regionalist tradition. In Sauer's rendition, this kind of geography subtly reversed the causal arrows so that the emphasis was on human agency and less on 'natural necessity'. This change of emphasis showed how suspicious of making causal links from environment to 'human nature' and society many geographers had become by 1939. Finally, there was the very tentative emergence of human geography studied in relative (but not absolute) isolation from environmental issues.

It seems to me that, almost half a century after the discipline was first professionalised, nature remained central to geography's identity – but also to its internal problems and its persistently precarious status within academia. Its centrality to geography's identity should be clear enough from the discussion above, even after the excesses of environmental determinism had been tempered. Geography was less of a naturalistic subject by the 1930s than it had been some decades before, but it was still highly preoccupied with nature as an object of analysis and a causal force in its own right. As for the discipline's internal problems, they revolved around how best to study nature, just as they had fifty years before. Three problems loomed large. First, the declining popularity of evolutionary thinking within geography was both a blessing and a curse. On the plus side, it took discussions of human nature out of most geographers' purview and, as noted above, it tempered the crude causal statements and moral judgements often characteristic of determinist discourse. Discussions of human psychology were being revolutionised by Sigmund Freud's ideas and, as the discipline of human biology grew larger, it became clear that geographers had little of substance to say about the human mind and body. On the downside, though, the loss of an overarching theory tying people and environment together was a costly one. It left geographers without a coherent conception of causality that would 'bridge' the social and natural sciences. It also meant that they increasingly became empiricists: that is, compilers of facts and describers of natural and human landscapes. This was especially evident in the numerous regional monographs that became pre-war geographers' stock-in-trade. At their worst, these monographs simply listed the various human and physical characteristics of a region in chapter after chapter. At their best, they offered a creative interpretation of what made a region unique. Fleure's (1926) *Wales and Her People* was typical of the genre. In elegant prose, Fleure evoked the 'spirit' of Wales – that peculiar combination of Celtic history, Anglo-Saxon invasion and rugged, maritime environment. His book was more an exercise in interpretative, impressionist analysis than scientific rigour.

Regional monographs also encapsulated geography's second internal problem – one that again was intimately connected to the question of nature. As the mid-twentieth century approached it was clear that geographers simply studied too much. The problem bequeathed by geography's founders simply wouldn't go away. Despite the attempts of some, like Sauer and Davis, to focus on one 'side' of the people–environment dialectic, geography was still saddled with a hopelessly broad subject matter. This reinforced the above-mentioned empiricism of the discipline. If, by the 1930s, geography was the 'integrative' subject Mackinder and Herbertson had wanted it to be, then it was largely in the descriptive sense and little more. Description matters of course, especially when it involves representing unknown things or familiar things cast in a new light. But explanation matters too and yet the explanations offered by geographers often did not convince.

Of course, there was still a possibility of making geography a convincingly causal subject, not such much by narrowing the focus of geography but by narrowing that of individual geographers. As already noted, Davis's denudation chronology and Chisholm and Smith's economic geography, showed that a division of labour was possible within the discipline. Physical geographers could focus on the different elements of the natural environment and their interconnections, leaving other geographers to focus on the human dimension. The sum of these individual specialisms would still, perhaps, make good on geography's holist ambitions. But founding 'half' the discipline on the study of the environment alone opened the door for geologists, botanists, zoologists and others to colonise the physical geographers' desired intellectual territory. Likewise, creating a human 'half' of geography begged the question of how that half was to distinguish itself from the social-sciences and humanities disciplines. This was pre-war geography's third internal problem – one, again, inherited from the subject's early history.

Externally, what all this meant was that other disciplines often had two negative views of geography by the 1930s. As Herbst (1961: 541) later noted, geographers 'suffered from the dubious reputation of being interlopers and second-rate performers in the fields of geology, meteorology, geophysics and plants and animal ecology . . . and pseudo-sociologists, pseudo-political scientists, economists and historians'. Second, the breadth of regional geographers' interests and their often impressionistic accounts of regions gave the discipline 'a dilettantish image among the practitioners of ever more specialising sciences' (Livingstone 1992: 311). As R.J. Russell reflected in 1949: 'I could not escape the conclusion that the position of geographers is not one of high esteem. I found the field criticised sharply on all sides'. In a retrospective, Peter Gould (1979: 140–1) even went as far as to say that 'it was practically impossible to find a book in [pre-1940s geography] . . . that one could put in the hands of a scholar in another discipline without feeling ashamed'. This dual image problem had especially dire consequences in the USA. In 1948 the President of Harvard, James Conant, opined that 'geography is not a university subject' (Smith 1987: 159) and the university's geography department was closed shortly thereafter. More generally, American physical geography had, by the mid-twentieth century, failed to win the battle for independence with geology and other disciplines (Leighly 1955).

POST-WAR RUCTIONS

Two geographies?

After the Second World War the nature of geography swiftly altered. Post-1945 there was a generation of new and established geographers who had lived through one of the most destructive wars in human history and who, especially in Europe, inhabited societies in desperate need of physical reconstruction and economic revival. Many of these geographers had worked in the military and intelligence services during the war years. Their expertise in cartography, land-use inventory, resource classification and regional taxonomy was useful in everything from logistics to battle planning. But by the late 1940s, geographers like Edward Ackerman (1945) came to the conclusion that their pre-war geographical education had failed them. The perceived failings were twofold. First, geographers lacked topical expertise, while their regional expertise was topically shallow. Second, wartime geographers lacked the technical and methodological skills to precisely measure real-world phenomena – a weakness when it came to meeting the demands of military and civil planning.

It was in this context that a post-war cohort of geographers set about reinventing the discipline – but not before a University of Wisconsin geographer, Richard Hartshorne (1939), had sought to cling to the past. Hartshorne's enormously influential *The Nature of Geography* set about telling geographers just exactly what their discipline should be (as its stentorian title suggests). It was the most sophisticated attempt to define geography to date and drew, in a way Hartshorne's predecessors had not, upon a lofty philosophical literature to justify its arguments. In keeping with the pre-war regionalists, Hartshorne defined geography as the study of 'areal differentiation'. Geography, for him, was the study of the unique and particular, whereas most other disciplines examine general patterns and processes. Downplaying the human–environment theme (no doubt because of environmental determinism's poor credentials), he strategically emphasised the conjunction of phenomena in different places as geography's key focus (cf. Entrikin 1981).

However, Hartshorne's synthetic sensibilities were soon challenged by others seeking to beef-up the emerging 'systematic' (i.e. topical) branches of geography at the expense of the regionalists. Among the key contributions here were those of Bagnold, Horton and Strahler, on the physical side, and that of Fredric Schaefer in relation to both human and physical geography. What these contributions did was accentuate the mild pre-war tendencies towards topical specialisation in geography. More than this, they also sought to make geography not the study of regions or human–environment relations but, rather, a *spatial science*. This was clearest in Schaefer's (1953) attack on Hartshorne. An economist by training and a German émigré who'd been interned by the Nazis, Schaefer was influenced by the Vienna School in the 1930s. This school sought to establish what science is. Finding himself in Iowa University's geography department after the war, Schaefer believed that geography could be a science in the same way that physics or chemistry were sciences. But what was a science? And what would distinguish geography from other sciences? Schaefer's answer was that all science is based on careful empirical observation, has explanation as its goal and its ultimate quest is the identification of general laws that underlie the behaviour of all sorts of different phenomena (like the law of gravity). In the case of geography, Schaefer saw its role as explaining the spatial patterning of human and physical phenomena. As he put it: 'Geography has to be conceived as the science concerned with the formulation of the laws governing the spatial distribution of certain features on the earth's surface' (1953: 227). Geography was thus to be defined, once more, not by its subject matter – which it shared with other disciplines – but by its perspective (the spatial distribution of things) (see Box 2.2).

In the same spirit as Schaefer's intervention, several other critics with environmental interests paved the way for a physical geography based on precise measurement and whose goal was the identification of the general processes producing landforms, water courses, soil profiles, vegetative communities and climatic and weather patterns. Bagnold's (1941) *The Physics of Blown Sand and Desert Dunes* inquired into process-form connections in arid environments (Bagnold's British military service had been in dryland regions). R.E. Horton (1945) used his engineering background to argue that the action of water over and through different types of soil and rock had consistent physical consequences that could be empirically measured – and even predicted. Finally, Strahler's (1952) 'Dynamic basis of

Box 2.2 GEOGRAPHY AS A SCIENCE

The attempt of Schaefer and others to make geography a 'spatial science' after the Second World War can convey the false impression that pre-war geography considered itself to be 'unscientific'. In reality, the founders of anglophone university geography (like Mackinder) certainly saw the subject as a science, as did Schaefer's erstwhile opponent, Richard Hartshorne. This begs the question: what is science? There is no single nor correct answer to this question. For instance, the *Oxford English Dictionary* offers four main definitions of the term. In the late Victorian period, anglophone geography's founders were arguably working with a very generic understanding of science. They saw it as an attempt to understand the world through systematic observation of the human and non-human worlds. In this sense, they distinguished science from opinion, religion, metaphysics, dogma and mysticism. Science, for them, was evidence-based knowledge and, as such, was a potentially objective reflection of reality: that is, about facts not fiction. However, during the early twentieth century, most geographers rarely got beyond this very general, rather 'thin' conception of what made science different from other human practices and scientific knowledge different from other ways of knowing. Schaefer's paper and the subsequent efforts of Bunge (1962), David Harvey (1969) and other 'spatial scientists' were attempts to offer a more specific or 'thick' conception of science. According to some this conception was 'positivist' or 'logical positivist'. I do not have the space to explain these two conceptions of science here, but I can make some general points. First, both take it as axiomatic that a material world exists independently of the investigator (the postulate of 'ontological realism') and that scientific knowledge can accurately mirror that world if appro-priate procedures are followed (the postulate of 'epistemological realism'). Second, both argue that the material world can act as a 'court of appeal' to adjudicate between rival interpretations of its true nature. Using our senses and various instruments, both maintain that we can ascertain empirical truths 'out there' in the world. Finally, both notions of science regard the use of systematic, repeatable investigative procedures as the best way to avoid bias and ensure that reality can, as it were, 'speak for itself'. For more on positivist ideas in human geography see Johnston (1986: ch. 2) and in physical geography see Inkpen (2004). I will say more about science in general in Chapter 4, en route to a discussion of physical geography's scientific credentials.

geomorphology' argued strongly that physical geographers should measure and explain how processes defined by universal laws create specific sorts of landforms given a certain set of initial conditions. Together, Bagnold's, Horton's and Strahler's works exposed the weaknesses of Davis's imprecise, non-quantitative and ultimately speculative theories of landform evolution and, by extension, of all earth-surface phenomena. They laid the groundwork for a physical geography where explanations were derived from the testing, by way of repeated observation and measurement, of refutable hypotheses.[1] This was an altogether more specialised, more rigorous, less descriptive approach to the physical environment than almost anything found in pre-1939 geography. It was aided by the development of new techniques (like pollen analysis and aerial photography) for measuring environmental phenomena in the field or in the laboratory. Within a decade key texts like *Fluvial Processes in Geomorphology* (Leopold et al. 1964) were making this new kind of physical geography a serious proposition.

Mirroring this, the 1950s and 1960s saw human geography emerge as a distinct part of the discipline with its own subfields. The idea was that, like physical geography, it could survive vacating the integrative space claimed by regionalism and human–environment studies by being a locational science. It would describe, explain and maybe even predict the spatial patterning of the phenomena studied ageographically by economists, sociologists and political scientists. Economic and urban geography made great strides in this regard. Young Turks like Brian Berry, William Bunge and William Garrison in the USA all insisted that there was a spatial order to economic and urban life and set about identifying it and the general processes that brought it into existence. More generally, Peter Haggett's (1965) landmark book *Locational Analysis in Human Geography* argued that all the phenomena human geographers might wish to study – from migration to transportation – have a non-random spatial configuration that can be explained in terms of a few key principles or processes. In this quest to be a distinct branch of the discipline, human geography in the USA was much more successful than its physical counterpart, which never quite escaped the clutches of geology departments. In the UK and Commonwealth countries, though, both human and physical geography boomed during the 1950s and 1960s – years when governments invested heavily in their universities.

Though the differences between human and physical geography hardened through the 1960s, this did not, initially at least, mean that geography was a divided discipline. Despite the obvious differences in subject matter, human and physical geography could claim to have the following key things in common. First, there was the joint commitment to describing and explaining the spatial distribution and spatial patterning of things on the surface of the earth at various scales – from why so many river-tributary systems are dendritic to why migration volumes decline with distance from the migrants' source area. Second, both human and physical geography employed a similar investigative procedure, namely the deductive-nomological procedure (or what was known as 'the scientific method'). This procedure, however loosely followed in practice,[2] ensured that whatever their subject matter, all aspiring scientific geographers would investigate reality in a similar way. As David Harvey explained in his methodological treatise *Explanation in Geography* (1969), the scientific method entailed the following steps. To begin with, a researcher would carefully observe a portion of reality that interested them and would then seek to explain what they saw in terms of clearly articulated hypotheses. These hypotheses would then be tested to see if they were confirmed by numerous attempts to verify and/or falsify them empirically. In turn, once the empirical evidence was sufficiently voluminous, a theory or law would be derived from the now-substantiated hypotheses that would apply to all other instances of the phenomena they covered not yet studied. This meant that, in future cases, one might predict a set of events given one's faith in the law or theory and enough local knowledge about the specifics (or 'initial conditions') of the case in question.[3] This mention of laws and theories brings us to the third thing human and physical geography had in common during the 1950s and 1960s: namely, a commitment to discovering laws and developing theories (and also models) that were of a wide applicability within various sub-disciplinary areas. In economic geography these included location theory and in population geography the 'gravity law' of migration. Titles of new books – like Bunge's (1962) *Theoretical Geography* and Scheidegger's (1961) *Theoretical Geomorphology* – boldly announced geography's new emphasis. Finally, during the two decades after the Second World War, those geographers aspiring to be scientific specialists tended to favour numerical measurement and the use of descriptive and inferential statistics in their data collection and analysis. Indeed, one geographer in the 1960s felt it no exaggeration to talk about a 'quantitative revolution' in the subject after 1945 (Burton 1963).

The shrinking centre

While geography's traditional subject matter was being apportioned to one or other 'side' of the discipline, other geographers still wished to occupy the middle ground so beloved of their early twentieth-century forebears. Even during the 'spatial science' revolution there were many geographers continuing to work in the regional mould commended by Hartshorne or pursuing the cultural landscape research advocated by Sauer. Some of these geographers came to see regional study as an 'art' – an exercise in interpretative and imaginative synthesis (e.g. Gilbert 1960). But others, relatively small in number, sought to reinvent the human–environment tradition of research and teaching. Chief among them was Gilbert White at the University of Chicago. White was a student of Barrows. His *Human Adjustment to Floods* (1945) helped inspire a renewal – by way of narrowing their focus – of human–environment studies in geography. This narrowing entailed, first, a concentration on how people adjust (or fail to adjust) their behaviour in relation to extreme physical events (like floods). Second, White was interested in the human side of this equation: in how peoples' *perceptions* of hazard risk affected their choice of where to live, where to work, and how to reduce their vulnerability to geophysical threats (see also Saarinen 1966). White's work inspired his students – Ian Burton and Robert Kates – to undertake applied research in flood management. This research developed policies sensitive to how people's often idiosyncratic perceptions of hazards affected their decision-making and hence their actual vulnerability to those hazards. Managing floods, White and his students showed, was not just about physical planning but also about understanding people's mental maps of the world.

The topically specific, empirically based and policy-relevant research of these geographers – whatever its other merits – did not provide much of a bridge between geography's two fast-growing halves during the 1950s and 1960s. For others in the discipline something more encompassing was needed to hold geography together – not a causal theory, like evolutionism, but something still concrete enough to act as a glue. That something was systems theory. Systems theory was more a useful analytical *vocabulary* for studying all sorts of different things than a *theory* in the normal sense of that term. It had diverse origins outside geography in Tansley's 'ecosystems' thinking and von Bertalanffy's general systems thinking. In geography, Chorley and Kennedy's (1971) *Physical Geography: A Systems Approach* was the first programmatic statement. Although this book presented systems theory as a way of bringing physical geography's growing subfields together, it had a wider relevance. Systems, Chorley and Kennedy showed, comprised elements, relationships between elements (simple or complex), and inputs and outputs (of energy and matter, for instance). They showed that systems concepts like 'homeostasis', 'negative feedback' and 'positive feedback' could be applied to all manner of topics. And they usefully distinguished different types of system (open, closed, cascading, process-response and so on). In relation to human–environment relations, people could be seen as one element of often-complex systems where human and physical component interacted in patterned ways with identifiable consequences (see Bennett and Chorley 1978).

In a sense, systems theory offered the sort of integrative promise evolutionary theory had decades before – but minus the spurious causal claims and replete with a scientific-sounding vocabulary. Strangely, it never really caught on in the discipline as a whole – despite some useful empirical work by Bernard Nietschmann (1973), among others. Instead, it became the framework of choice within the physical geographic community (and is still prevalent in that community today, see Gregory 2000: ch. 4; Inkpen 2004: ch. 6). But even if systems thinking had caught on in geography as a whole, the fact that it was really a nomenclature – a descriptive device that offered a common conceptual language for human and physical geography alike – meant that its precise operationalisation was always going to differ from geographer to geographer. At best, systems theory would have been a weak glue holding geography's emergent siblings together. The tragedy, of course, is that geography's failure to prevent a progressive split between human and physical geography came at a time when the

Western environmental movement first gathered momentum. By the late 1960s it was clear that population increase, economic growth and mass consumption were having a profound effect on natural resource availability and the integrity of ecosystems. Mercury poisoning at Minamata Bay, Japan; the Torrey Canyon oil-tanker spill; Rachel Carson's (1962) best-selling account of how herbicides and pesticides got into the food chain: these and other incidents inspired the first Earth Day, the founding of Greenpeace and Friends of the Earth, and other seminal early-1970s environmental initiatives. Geography had a golden opportunity to make 'human impact' studies its main business – a possibility foreseen in 1956 in *Man's Role in Changing the Face of the Earth*, in which an ageing Carl Sauer lamented the environmental degradation wrought by industrial societies.

Quite why this opportunity was missed is hard to say (see Simmons 1990 for speculations). Although it was grasped at the teaching level, this was not the case at the research level. Chorley's (1969) *Water, Earth and Man* – which called for a new focus on human–environment interactions – was arguably the exception that proved the rule. Geographers conspicuously failed to analyse the local and global 'environmental problems' that became ever more apparent from the early 1960s (Mikesell 1974). Morally, the discipline also virtually ignored the pro-nature (or ecocentric) arguments being made within the wider environmental movement. Instead, a relatively small number of geographers complemented the natural hazards work of White et al. with a rather anthropocentric focus on resource management. This resource analysis was usually empirical, quantitative and conducted in what geographer Tim O'Riordan (1976) called a 'technocentric' mode. In other words, this research looked at how best to conserve resources for present and future human needs. It rarely took issue with the fundamental causes of resource depletion and was very human-centred (Box 2.3). Relatedly, a number of physical geographers were interested in the impact of human activities on the parts of the environment that interested them (and vice versa) (e.g. Hollis 1975). Like resource-management studies, their research had a policy dimension because environmental management needed to be based on a proper understanding of its objects (e.g. rivers, soil erosion, predator–prey relationships). Yet despite its apparent 'ethical-neutrality', this sort of research was arguably value-laden because it did little to challenge the human actions and value systems that generated environmental degradation in the first place. It was very much 'status quo' research.

Box 2.3 ATTITUDES TOWARDS THE NATURAL ENVIRONMENT

According to O'Riordan (1989), *ecocentrism* and *technocentrism* are the two dominant attitudes towards the non-human world in Western societies. The former is a pro-nature attitude that has mild and radical variants. The mild variant ('communalism') recommends a return to small-scale communities using local environments in a sustainable way and using clean technologies. The radical variant (sometimes called a deep or dark green attitude) suggests that the non-human world has inherent rights that should be respected. Technocentrism, in contrast to eco-centrism, sees the non-human world as a means to the end of human well-being. In its mild 'accomodationist' form, it puts faith in the adaptability of technology and institutions so that when resource scarcity or environmental problems arise they can be adjusted to without a decline in living standards. In its more radical 'interventionist' form, technocentrism puts faith in the power of technology and ingenuity to transform the environment for human well-being – as exemplified by genetically modified foods. The mild forms of both ecocentrism and technocentrism do not pose a fundamental challenge to the way Western societies currently organise their use of the environment. By contrast, the radical strands do, which is arguably why they are less popular. Deep greens call for nothing less than a new environmental ethic that puts people and the non-human world on a level moral pegging. Meanwhile, left-wing interventionists – like several Marxist theorists – want a post-capitalist future in which everyone can enjoy a high standard of living not just the wealthy few. In this post-capitalist future the exploitation of the environment would meet the general needs of the population. Technocentrism – in both its accommodationist and interventionist forms – is *anthropocentric*. That is, it puts people ahead of the environment. The spectrum of environmental attitudes described above can be found in microcosm vis-à-vis animals. For instance, extreme animal-rights activists believe that animals have rights just as humans do (see Wise 2000 for a sophisticated case), while interventionist-technocentrics are happy to genetically modify animals for research and nutritional purposes.

Knowledges of nature

As the previous two subsections make clear, there'd been something of a sea change in geography's approach to understanding nature by the late 1960s. First, the 'nature' in question was almost exclusively the non-human world, biotic and abiotic, animate and inanimate. Second, the study of this nature became the preserve of physical geographers, a smallish cohort of human–environment geographers and a smattering of regional geographers. Third, physical geographers were also interested in nature in the sense of both 'essence' and 'inherent force'. Their agenda was to disclose the true character of environmental processes and the effects they had on and at the earth's surface. This was greatly facilitated by the tendency of these geographers to favour highly empirical, small-scale, case-study research based on careful field and laboratory analysis. Finally, the relatively young but fast-growing field of human geography was largely non-naturalistic by the late 1960s. Its subject matter was people and the spatial organisation of their activities. Human geographers during the spatial-science years avoided talk of 'human nature' – except in those models and theories that assumed universal human characteristics (like 'rationality' and the desire to 'minimise effort') in order to generate testable hypotheses about real-world spatial patterns. This was the start of a four-decade process of making human geography a social science as distinct from a natural one, as well as a humanities subject too.

How do we explain this shift in the nature geographers studied, in the geographers who studied it and in how they studied it? The aspiration to make physical geography a science had obvious appeal during the mid-twentieth century. Science, after all, had the image of being more rigorous than any other knowledge-producing activity. By employing precise measurement techniques, the scientific method, and theories, laws and models, physical geographers could gain respectability as analysts of the non-human world. They were to be field scientists, paralleling the laboratory scientists and applying some of the latter's insights (see Chapter 4). Meanwhile, if human geography was to study the spatial patterning of what the social sciences and humanities studied ageographically, then it could ill afford to study the environment too. To place this in context, it's important to note that in the West the post-war state apparatus increased in size enormously. National governments played a major role in public life after 1945, not least through elaborate welfare-state provisions for their citizens. They took on management functions in relation not only to society but to the environment also (for instance, the US Environmental Protection Agency was created by President Nixon in the early 1970s). This created a niche for both human and physical geographers, as well as those occupying the zone between the two. For instance, many human geographers became actively involved in transportation policy, urban and regional planning, and industrial location policy. They could use their models and theories to analyse and plan the spatial organisation of societies. Physical geographers were likewise able to use their scientific credentials to attract state funding for 'pure environmental research'. Finally, human–environment geographers could contribute to resource and environmental management wearing their fact-based, technocratic hats.

By the early 1970s, then, the knowledges of environment being produced by geographers possessed the following characteristics. First, in keeping with the scientific paradigm that had swept through geography, much of this knowledge was claimed to be *realistic*. In other words, it was knowledge about a non-human world whose characteristics, it was believed, could be accurately described, explained and possibly predicted. This sort of realistic knowledge was contrasted sharply with the sometimes spurious assertions of the determinists of an earlier generation and with the evocative prose of many regional geographers. Second, geographers' knowledges of environment at this time were largely *cognitive* ones. Questions of morality, ethics and aesthetics vis-à-vis the non-human world were left largely to philosophers, poets, and historians of ideas. Third, this was linked to a belief that geographers' research on the environment was *value-free* for the most part. This meant not only that researchers bracketed their own values so that the facts of nature could 'speak for themselves'. It also meant that the knowledges of environment that geographers produced were believed to be 'neutral'. Values were supposed to be exterior to scientific research not bound into it. Finally, a good deal of geographers' research about the environment at this time was intended to be *instrumental*.

Put differently, it was geared to the control and prediction of environmental phenomena.

On this basis, physical and human–environment geographers attracted large amounts of public research funding from the early 1960s. As universities expanded, the numbers of undergraduate and graduate students increased significantly. These students were attracted by the idea of receiving a rigorous 'scientific' education that would equip them for the world of employment. For those students with strong environmental interests, a geographical education at this time would give them specialist knowledge in the workings and husbandry of everything from coastal environments to grassland ecosystems to fish resources. This was just the sort of knowledge that local and central governments needed in the discharge of their environmental-management duties. It was also the sort of knowledge that firms whose business was resource exploitation – like British Petroleum plc – needed in their workforce. Of course, there were exceptions. Many geography departments in the English-speaking world perpetuated pre-war traditions of research and teaching, while others saw human geography less as a spatial science and more as the 'art' to which I referred earlier. In the USA, for instance, Sauer's influence lived on, while rising stars like Andrew Clark (of Wisconsin University) created related approaches to cultural-historical landscapes. Meanwhile, in the UK, the study of 'human regions' was perpetuated in geography departments like Aberystwyth.

ONTOLOGICAL DIVISION AND THE DE-NATURALISATION OF HUMAN GEOGRAPHY

As the 1960s gave way to the 1970s, the differences between human and physical geography grew. It was not simply that the former studied people and the latter studied the environment. It was also that the *way* they studied their respective subject matters began to alter. As I'll explain in this section, the *ontological differences* between people and the non-human world became the foundation upon which geography's two halves began to travel in different directions. At the same time, as I'll also explain, several human–environment geographers began to de-emphasise the 'naturalness' of the environmental side of the relationship that interested them.

A nature-free human geography?

One of the problems with the theories and models put forward by 'scientific' human geographers was that they offered highly approximate descriptions and explanations of migration patterns, industrial location, commuting behaviour and the like. Among other reasons, this was because they made simplistic assumptions about how human actors make decisions in real-world contexts. In effect, 1960s human geography was 'in-human': it failed to understand the complexities of real people living and acting in concrete situations. Instead, there was a preference for analysing large data sets about the number and destination of migrants; about the number, type, and location of industries; about the volume and distance-decay charac-teristics of commuting; and so on. In short, 'spatial scientists' on the human side of geography studied people 'at a distance'.

This paved the way for what became known as behavioural geography. Among its precursors was the hazard-perception work of White, Saarinen and others. Formally inaugurated with Cox and Golledge's (1969) *Behavioural Problems in Geography*, this approach 'promised the construction of more realistic and human-centred models of the world' by 'focusing on the complex ways that people obtain sensory information from, make sense of, and remember their surroundings' (Hubbard et al. 2002: 36). Ontologically, behaviourists argued that people are not the same as rocks or atoms (echoing Dilthey's late-nineteenth-century arguments). If there's an order and regularity to human decision-making and human action it is, they argued, a 'fuzzy' one. On the basis of this belief, behavioural geographers examined how different people process information from their surroundings, mould it into definite thoughts, beliefs and attitudes, and then undertake actions on this basis. Such an approach took it as axiomatic that if human decision-making could be described in law-like terms then these terms would be stochastic and probabilistic not rigidly deterministic. The theories and models used came from psychology, landscape-planning and micro-sociology. In methodological terms, behavioural geographers measured people's perceptions, understandings and attitudes using psychometric tests, questionnaires, rating scales and the drawing of mental maps. Overall, behavioural geography challenged the universal rationality postulate of the spatial scientists. If people had a 'nature' at all, the behaviourists argued, it was their special capacity to think and act in context-specific ways.

As noted, behavioural geography resonated with earlier work by geographers – such as White's perception studies. But in its desire to quantify and measure the links between people's thoughts and their behaviour, and in its preoccupation with applying psychological theories, it was an altogether more systematic and ambitious exercise. It was, if you like, spatial science with a human face. It aimed for precise, realistic knowledge of spatial decision-making that was objective and value-free – only more precise and realistic than that offered by 1960s human geographers. But this kind of human geography was not to everyone's taste.

A cohort of 'humanistic geographers' writing from the early 1970s onwards, wanted to take things a step further. Echoing Dilthey's arguments more strenuously than the behavioural geographers, Yi-Fu Tuan, David Ley, Edward Relph and several other young researchers of the time argued that the assumptions and procedures of science were simply not appropriate for the study of people. Humans, they argued, are not just rational beings but also *moral* ones, not just thinking beings, but also *emotional* ones who possess feelings and desires. As one of their early advocates argued: 'Humanistic geographers [believe] . . . their approach deserves the appellation "humanistic" in that they study the aspects [of people] which are most distinctively "human": meaning, value, goals and purposes' (Entrikin 1976: 616).

This approach was a *hermeneutic* one. It involved an attempt to gain empathetic understandings of different people's 'life-worlds': that is, the frameworks of understanding, belief and value particular to them. This kind of life-world research was often focused on how individuals or small groups of people gained attachments to particular places and specific local environments. It emphasised the subjective dimensions of human existence over the brute objectivity of built and natural landscapes. Key publications included Tuan's (1974) *Topophilia: A Study of Environmental Perception, Attitudes and Values*, Relph's (1976) *Place and Placelessness* and Graham Rowles's (1978) *The Prisoners of Space?* Methodologically, humanistic geography pioneered the use of interviews, focus groups and ethnography in human geography. In philosophical terms, it drew inspiration from the anti-materialist thinking of the European fin-de-siècle 'romantics' (Husserl and Kierkegaard), from Martin Heidegger's philosophy and from Jean Paul Sartre's 'existentialism'.

These methodological and philosophical innovations were not without consequence for human geography as a whole. The legitimation of qualitative techniques further distanced human geography from its physical counterpart. It also questioned the empiricism, observational detachment and generalising impulses of both spatial science and behavioural geography. Humanistic geographers argued that one had to 'get inside' people's heads in order to tease out the invisible thoughts and feelings that produce visible actions. But because researchers are human too, this meant that all life world research involved a 'double hermeneutic' – researchers could only uncover the 'realities' of other people's life-worlds from the particular perspective of their own. Obviously, this posed a major challenge to the idea that one could construct general theories and models of human thought and behaviour.

In principle, humanistic geography was well equipped to link human geography with the burgeoning environmental movement of the early to mid-1970s. Its focus on values and on the way people create attachments to the non-human (as much as human) world, could have led to a 'green human geography'. This geography might have studied how and why more and more people in the West were valuing the environment during the 1970s. It might have injected a 'pro-environment' (or ecocentric) morality into geography. And it might also have highlighted ordinary people's aesthetic appreciation of the environment. But this never happened (despite Tuan's [1974] pioneering study and Seamon and Mugerauer's [1985] later collection). Instead, humanistic geography was responsible for reintroducing another 'nature' into geography. This was the idea of 'human nature', albeit in a very abstract, non-biological form that made no reference to the physical environment or to specific physiological or psychological processes. For humanistic geography, what *all* people had in common was their innate capacity for complex and changeable thought and feeling. This very general claim about 'human nature' had a normative dimension: for what worried many humanistic geographers was that different people's life-worlds were being assaulted by a homogenising triad of consumerism, industrialism and governmental intervention that was eroding 'areal differentiation' (see Relph's [1976] highly moralistic study, for example). Their abstract conception of 'human nature' was thus an ethical weapon: humanistic geographers wanted to show that any uninvited encroachment on people's life-worlds was, *ipso facto*, a bad thing.

This brings us to a second new approach (or paradigm) within 1970s human geography that also took issue with spatial science and behaviourism but which departed from humanistic geography at the same time. This was

Marxist geography. During the late 1960s many people in the West expressed dissatisfaction with the post-war order. This extended beyond environmentalists' critiques of resource exploitation, species loss and the like. In addition, there was the civil rights movement in the USA, the anti-Vietnam war protests, the Algerian war of independence, the 'events' in Paris in 1968, and the challenge to capitalist societies posed by the communist bloc. On top of this, there were several epic famines in the developing world (something of a harbinger), and a concern that the West's affluence was being bought at others' expense. In this context, spatial science and its behaviourist offshoot seemed not only to ignore the most pressing issues of the day. They also appeared to be 'part of the problem' in so far as they failed to challenge imperialism, racism, oppression, poverty and other social ills in any meaningful way. As David Harvey put it in his landmark book *Social Justice and the* City: 'There is an ecological problem, an urban problem, an international trade problem, and yet we seem incapable of saying anything of depth or profundity about any of them . . . The objective social conditions . . . explain . . . the necessity for a revolution in geographic thought' (Harvey 1973: 129).

One can speculate why Harvey and his students turned to Marxism, rather than to any other critical theory of society, to engineer this 'radical geography' revolution. First, Marx's ideas, as was well known at the time, had inspired the supposedly 'emancipatory' experiments in communism in the USSR, eastern Europe, China, Cuba and several other countries. I say supposedly because Marx's ideals – unbeknownst to many in the West – were being perverted by dictatorial communist leaders. Unaware of this, many left-wing individuals in the West saw communism as a living, humane alternative to capitalism and Western economic imperialism. Second, academic Marxism – that is, the critical analysis of how capitalism works – was very influential in Western sociology, anthropology and philosophy departments during the 1960s. This gave Harvey and his students a tradition of thought ready to hand, as it were. Finally, because Marxism's main object of analysis was capitalism, and because capitalism was an increasingly global economic system, Marxism's insights seemed to have wide relevance and applicability.

This is not the place to rehearse the history of Marxist geography. For our purposes it's enough to note the following. First, like humanistic geography, the Marxist approach said little about the environment. This was arguably a reflection of Marx's relative inattention to the topic (an

inattention Marxists have rectified only in the past fifteen years or so).
Second, Harvey and other geographical Marxists mostly eschewed any
talk of 'human nature'. For them, Marxism was a *historically specific theory* of
the particular *social relationships* specific to capitalism. Marxist geographers
were mostly interested in the 'structures' (economic largely, but also social
and political) that explain why some people in some places enjoy wealth
and prosperity while people in the same places and elsewhere suffer
poverty, unemployment and malnutrition. In other words, Marxists rejected
the individualistic and small-group focus of behavioural and humanistic
geography. Instead, they paid attention to the *unequal power relations* between
connected social classes (like employers and workers). They challenged
the sociologically 'thin' conception of the human person offered by the
humanists and emphasised that people's thought and action can be properly
understood only within a 'thick' conception of how any specific society
operates.

Marxist geography helped to expunge nature from human geography
during the 1970s (human and non-human). So did behavioural and
humanistic geography, despite the abstract claims concerning human nature
characteristic of the latter. All three approaches extended the subject matter
that geographers took a de-naturalising approach to. Everything one needed
to know about individuals, groups and social structures was to be found
in habits of thought, interpersonal relationships, cultural norms and
so on. People on the ground were examined *contextually* in these three
approaches, not in terms of some 'fixed' internal or external nature to which
they supposedly conformed. At the same time, nature in the sense of
the environment didn't figure topically in much of the research of the
behaviourists, humanists and Marxists. The knowledge produced by the
latter two was intended to offer an alternative conception of the 'humans'
in human geography when compared to the scientific world view. These
knowledges – interpretative–hermeneutic and critical–emancipatory
respectively – were a challenge to the instrumental–technical knowledge
produced by the spatial scientists and behaviourists. Humanists and Marxists
saw people as ends in themselves, not as objects to be managed or means
to the end of others' wealth.

De-emphasising the environment: unnatural hazards and Third World political ecology

As human geography was progressively de-naturalised, something similar began to happen in the disciplinary middle ground. The natural-hazards research and teaching tradition inaugurated by Gilbert White came in for heavy criticism during the 1970s. A key publication was *Interpretations of Calamity* (Hewitt 1983). This influential volume argued that natural hazards were less 'natural' than meets the eye. Hewitt, the book's editor, didn't deny that floods, earthquakes and tsunamis were natural events. Rather, he questioned the way the way the hazard–human-response link was studied by geographers. By the 1970s more people than probably any point in human history were dying at the hands of natural hazards. Yet there was no evidence that the physical environment was any more or less capricious than in earlier decades and centuries. The White tradition of research would've explained this increased mortality with reference to individuals' perceptions of their vulnerability. In terms of hazard mitigation, this tradition focused (on the human side) on 'correcting' misperceptions or zoning land so that people couldn't occupy it, or else (on the physical side) on engineering solutions to hazards (flood barriers, sandbags etc.). Hewitt, by contrast, argued that individuals' choices of where to live and what to do are *structured for them* by the particular social position they occupy. Specifically, the poor often suffer the brunt of 'natural hazards' and Hewitt argued that there was nothing natural about this. He called for a 'social' approach to hazard analysis and mitigation that was less about the physical threats and more about who was *made vulnerable* to hazards and why.

This sort of critical-hazards analysis was complemented by the emergence of 'Third World political ecology' from the early 1980s. As the name suggests, this 'combine[d] the concerns of ecology and a broadly defined political economy. Together, this encompasses the constantly shifting dialectic between society and land-based resources, and also within classes and groups within society itself' (Blaikie and Brookfield 1987: 17). Despite the reference to land-based resources, Third World political ecology (TWPE) was more interested in the social side of the dialectic referred to from the start. In this it converged with Hewitt's agenda for hazard study. However, unlike Hewitt and his ilk, political ecologists were more interested in chronic environmental problems than the effects of extreme geophysical events. Empirically, TWPE focused on poor, rural

land-users in the developing world. Though the field included anthropologists, in geography TWPE emerged out of long-standing interests
in the non-Western world among regional geographers and out of a
dissatisfaction with the 'modernisation' theory of the 1950s and 1960s. This
theory had predicted that the developing world would follow the same
trajectory as the developed world. Yet, by the early 1970s, it was clear that
many developing countries remained land-based (non-urbanised), poor
and only partly industrialised. Drawing on political economy (a cluster of
left-wing economic theories of which Marxism is one), TWPE inquired into
how uses of land and resources at the local level were conditioned by a
hierarchy of social forces extending up the global level. For instance,
Blaikie's (1985) pioneering *The Political Economy of Soil Erosion in Developing
Countries* constructed a bottom-up analysis from individual land-use decisions (e.g. what crops to grow, whether to pay for irrigation measures, etc.)
to larger scales, at each level 'examining the social relations that shape
opportunities and constraints for land users' (Zimmerer 1996: 177). In this
regard, political ecologists shared Marxist geographers' preoccupation with
power relations and large-scale societal structures – indeed, many early
political ecologists were Marxists (like Michael Watts, whose research
I'll discuss in the next chapter). Politically, they wished to alter those
relations and structures so that poor, developing-world farmers would not
be forced into degrading the resources and environments upon which
they depended for their livelihoods. The close connections between
TWPE and critical-hazards analysis were later illustrated by the fact that Piers
Blaikie co-authored the important book *At Risk: Natural Hazards, People's
Vulnerability, and Disasters* (Blaikie et al. 1994).

The main exceptions to the de-naturalising thrust of human–
environment geography during this period were resource geography,
Sauerian landscape geography and cultural ecology. As noted earlier,
the former subfield emerged in the late 1960s when many governments
and publics became concerned about the seemingly precipitous decline
in natural-resource availability worldwide. Alarmist books like *Blueprint
for Survival* (Goldsmith et al. 1972), *The Population Bomb* (Ehrlich 1970) and
The Limits to Growth (Meadows et al. 1972) predicted a Malthusian future
of 'overpopulation', where a finite resource base would limit the numbers
of people who can live on the planet. Others, though, were optimistic
that technological innovation and ingenuity would allow more people
to live longer at a higher standard than ever before. In this context, the

United Nations held its first conference on population, environment and development in Stockholm (1972), while many Western nations pondered a future where their fossil-fuel supplies would have to be sourced from developing-world countries rather than their own shores. All this set the scene for the steady growth of a resource geography focused on assessing the quantity, distribution and availability of various renewable and non-renewable natural resources (see, for example, Mitchell 1979). Meanwhile, Sauer's students continued to examine how cultural groups transformed the natural environment over time. This cultural-landscape research shaded into cultural ecology, the holistic study of how different culture groups use and adapt to their local and regional environments (see Braun 2004: 153–9).[4] Figure 2.1 summarises the changing ways that geographers had studied society–environment relationship up until the late 1980s. Implicit

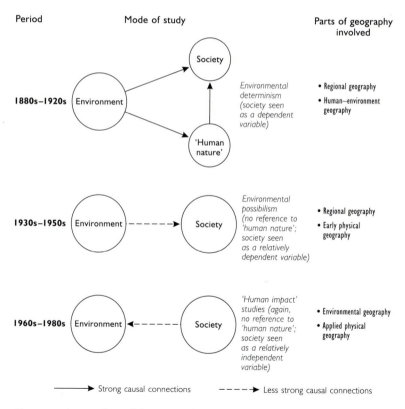

Figure 2.1 Geography and the study of nature–society relations, 1880–1990

in Figure 2.1 is the fact that as the decades wore on more and more geographers chose not to study this relationship, preferring instead to become specialist human or physical researchers and teachers.

Physical geography: pure and applied nature knowledge

While human geography was being de-naturalised, and the human–environment tradition was leaning to the people side of the human–environment relationship, physical geography continued the trend towards specialisation, case-study research, application of the scientific method, and the search for (and application of) empirically testable laws, theories and models. Geomorphology remained by far the largest subfield, within which various branches began to take definite shape (coastal geomorphology, glacial geomorphology, periglacial geomorphology etc.). A particular preoccupation with process–form relationships became evident (and endures to this day). This preoccupation had been inspired by key publications like Schumm and Lichty's (1965) paper on different spatio-temporal scales of analysis and Leopold, Wolman and Miller's (1964) pioneering *Fluvial Processes in Geomorphology*. In addition, new technologies like remote sensing and personal computing enabled physical geographers to measure and monitor the environment more extensively than ever before. The upshot was a stream of significant specialist publications about different aspects of the physical environment. These publications were both 'pure' and 'applied' in nature, the latter providing a physical counterbalance to the human emphasis of 1970s hazards analysis and TWPE. For example, in the subfields of hydrology and fluvial geomorphology there was Gregory and Walling's (1973) *Drainage Basin: Process and Form* and *Fluvial Processes in Instrumented Catchments* (1974), plus *Man's Impact on the Hydrological Cycle in the UK* (Hollis 1979) and Burt and Walling's (1984) *Catchment Experiments in Fluvial Geomorphology*. In these and other publications physical geographers evinced a faith that the environmental knowledge they were producing was realistic: that is, a faith that it could ultimately 'reflect' nature as in a mirror so long as rigorous investigative procedures were used. What's more, this environmental knowledge continued to separate statements of fact about the non-human world, from statements of value (moral or aesthetic).

This is not to say that physical geographers in this period all became specialists at the expense of any shared perspective on how the physical environment is structured across space and through time. As I noted earlier, systems theory – with its emphasis on interconnections, equilibrium, feed-back and mutual adjustment among system components – constituted an appealing language with which to analyse the connections within and between parts of the non-human world. As the 1970s wore on, the systems vocabulary was modified to accommodate the fact that physical environ-ments are often less orderly and stable than was sometimes supposed (Brunsden and Thornes 1979). Overall, though, the 1970s saw physical geography's subfields mature considerably in terms of new theories and models, new measurement and monitoring techniques, and large new data sets to analyse.

THE RETURN OF THE REPRESSED?

Human geography in the 1980s: the further erasure of nature

By the early 1980s human and physical geography had become relative strangers. It was not only subject matter that divided them (the one half of the discipline concerned with the natural world, the other half the human world): it was also styles of analysis. While physical geography remained broadly scientific in its approach, human geography increasingly became 'post-scientific' (or, to use a more technical term, post-positivist – positivism being a particular, once widely held conception of science whose precise influence on human geography remains disputed). Aside from humanistic and Marxist approaches, feminism also had a major impact on human geography from the mid-1980s. This came in the wake of a two-decade-old women's movement (especially vigorous in Europe and the USA), as well as dissatisfaction with the class issues (over)emphasised by the Marxists, and a dissatisfaction with the avoidance of social inequalities common in humanistic geography. Among other key publications, *Geography and Gender* (Women and Geography Study Group 1984) helped inspire a new generation of geographers to examine how patriarchy (the oppression of women by men) was reproduced and challenged in the various different physical and symbolic landscapes of modern societies – from the home to the workplace. In this first phase feminist geography tended be either 'liberal' or 'socialist' in outlook – the former seeking greater recognition for women within the confines of existing societal laws and norms, the latter offering a more radical critique of women's marginalisation within contemporary capitalist societies. Socialist feminists noted that many women suffered the twin oppressions of gender discrimination and class discrimination in the workplace, while having their domestic labour undervalued by men. They argued that only a dismantling of capitalism and patriarchal attitudes and practices would liberate women.

As feminist geography became more influential, other geographers were attempting a rapprochement between the structural–social relations perspective of Marxist geography and the free-will–individualist perspective of the humanistic geographers. This rapprochement was inspired by the work of sociologists like Anthony Giddens, whose 'structuration theory' provided concepts that promised to overcome the dualism between 'structure and agency'. In human geography, figures like Derek Gregory, Allan Pred and Nigel Thrift all adapted Giddens's work to show how the actions of people in specific places were conditioned by social forces operating at much larger geographical scales. This chimed with the aspirations of TWPE (though TWPE rarely used Giddens's ideas). Overall, the 1980s erasure of nature from human geographers' research was in keeping with the broader thrust of social science. Following Peter Winch's (1958) *The Idea of a Social Science*, sociologists, political scientists and anthropologists all gradually left the study of nature to the physical, medical, engineering and behavioural sciences.

Arguably, the only intellectual development of the 1980s that might have held human and physical geographers closer together was 'transcendental realism'. This ungainly name refers to an overarching philosophy pioneered by Roy Bhaskar and Rom Harre. Developed from the 1970s onwards, transcendental realism was a critique of conventional understandings of science (including positivism). It was, in essence, an attempt to explain to researchers in all disciplines the nature of reality (social and environmental) and how best to study it. Bhaskar and Harre believed that too many researchers operated with a flawed understanding of that which they studied. They took it as axiomatic that both societies and physical systems are *overdetermined*: that is, they are a complex, dynamic and not always stable amalgam of different causal powers. These causal powers are ones that are possessed by specific social and environmental phenomena by virtue of their internal structure (e.g. gunpowder has the power to explode because of its chemical composition) or the necessary relations they have with other phenomena (e.g. parents are normally responsible for their children because of law, custom and love). Quite how different causal powers interact is, Bhaskar and Harre insisted, a *contingent* question. The 'order' inherent in both social and environmental life, they argued, was not empirically observable but lay, rather, at the ontological level. The *same* causal powers could have different real-world effects depending on the context. In part, this is because of 'emergent effects' caused by the combination of two or more separate causal powers. What is more, because causal powers are not themselves visible their existence can only be inferred from observing these over-determined effects. This called into question the common belief, one that geography's spatial scientists had often adhered to, that the goal of scientific research is the identification of visible correspondences or patterns. For transcendental realists, the true aim of all research is to identify the enduring causal powers of things en route to an understanding of their contingent interactions in any situation, leading to equally contingent empirical outcomes.

Transcendental realism was introduced into geography by Andrew Sayer. His *Method in Social Science* (1984) explained realist ideas for human geographers. It helped consolidate the move away from spatial science in this half of the discipline and was sufficiently encompassing that Marxists, feminists and other human geographers could draw upon it. But the book also emphasised the ontological differences between the human world and the natural environment that was part of Bhaskar and Harre's philosophy. These differences included the fact that people are interpretative beings (unlike, say, trees) with a capacity to both reflect on and alter the social contexts in which their lives are lived — the point humanistic geographers had made from the early 1970s. Sayer's emphasis upon the implications of transcendental realism for social research no doubt allowed many physical geographers to ignore it until much later (in the 1990s), by which time human geographers had moved on to other intellectual pastures. Another possible reason why human and physical geography did not rally around transcendental realism is that physical geography's empirical, case-study, fieldwork emphasis made philosophical discussion less common than in human geography.

Whatever the reason, late 1980s human geography made virtually no reference to the environment, let alone notions of human nature, in any of its constituent parts. Theoretically, Marxism, feminism and the left-wing parts of humanistic geography paved the way for a 'critical human geography' that sought not only to *explain* the social world but to *change* it also. Topically, human geography was dominated by its economic, social and urban branches, with political geography growing in importance, development geography becoming more left-wing, and rural geography (including agriculture) left as a rather marginal subfield. Overall, human geography's strong engagement with social-science (and especially sociological) thinking during the 1980s cut it off from any concern with nature. Issues of the environment and 'human nature' were, in the main, left to other disciplines.

Human geography during the 1990s: the rediscovery of nature

Since the early 1990s human geographers have performed something of a volte-face regarding nature. For over a decade it has been 'on the agenda' in a way one would scarcely have anticipated in the 1980s. But it has (re)appeared in unconventional and unexpected ways. Let me explain. The United Nations' Earth Summit was symptomatic of a resurgence of environmental concern among governments, the public and even business during the 1990s. Unlike 1970s environmentalism – which often had resource exhaustion and 'overpopulation' as its major foci – 1990s environmentalism was preoccupied with anthropogenic environmental change. Global warming, ozone-layer thinning and 'acid rain' were just three examples of humanity's new-found power to create not merely local but also *global* environmental problems. Another difference from the 1970s wave of environmental concern was that, by the 1990s, philosophers, historians of ideas and political analysts had had two decades in which to fashion coherent moral doctrines for the conservation and preservation of the non-human world. Key figures like Arne Naess, Holmes Rolston III and Warwick Fox had helped to establish well-argued ecocentric positions that challenged the anthropocentrism of the ideas and practices typical of most societies worldwide. As a subset of this sophisticated 'nature-first' thinking the philosopher Peter Singer (and others) had made great strides in explaining why animals should have rights.

In this context, one might reasonably have expected a green human geography to emerge that was focused on people's attitudes towards, and uses of, the non-human world. As I observed earlier, such a geography did not emerge in the 1970s but could well have done two decades later. Morally, this green human geography might have been mildly or strongly 'pro-environment' and critical of environmental degradation in its various forms. Yet, in reality, 1990s human geography took a 'de-naturalising turn' that was focused on human as much as non-human nature. In other words, many human geographers have looked at those things that are often *thought to be natural* and argued that they are, in fact, wholly or partly *social, cultural and economic*. The philosopher Kate Soper (1995) has called this a 'nature-sceptical' stance. Simplifying, in relation to the environment this de-naturalising manoeuvre had two elements. First, some human geographers argued that *representations* of nature – whether held by environmentalists, ordinary people or anyone else – typically say more about those who advocate them than the 'nature' they supposedly depict. These representations might be verbal (e.g. in everyday speech), written (e.g. in newspaper articles) or visual (e.g. in wildlife documentaries or landscape art). An early example was Cosgrove and Daniel's (1988) path-breaking work on one of nature's 'collateral concepts', landscape. Taking landscape design and landscape painting as foci, they argued that power relations between different social groups found expression in the way landscapes were both physically arranged and subjectively viewed. They thus questioned the belief that landscapes were simply picturesque scenes or sources of sensory delight for all to enjoy equally. Another early example was Jacquie Burgess's (1992) research into a conflict over the use of Rainham Marshes, a conservation area near London. Burgess showed how Music Corporation of America and those opposing its planned development of the marshes both passed off highly specific and conflicting depictions of the area as ostensibly 'correct' ones.

Second, other human geographers examined the social relations, values and norms that led to certain *transformations* of the non-human world. This more material focus was not, however, inspired by a belief that certain environmental usages were 'anti-ecological' or 'unnatural'. Rather, the suggestion was that a good deal of the environment has not been 'natural' for a very long time. Indeed, the Marxist geographer Neil Smith (1984) felt compelled to talk about the physical *production of nature* in capitalist societies – a claim that now seems prescient in light of the 'biotechnology' revolution in agriculture, forestry and aquaculture. In sum, since the early 1990s many human geographers have shown that, in both representational and physical terms, the non-human world is in some measure a *social construction*.[5]

This de-naturalising focus on the non-human world may seem perverse given the apparently pressing environmental problems that now confront humanity. In questioning the 'naturalness' of an external nature it may seem to undermine the claims of environmentalists and to hold out little hope for a green geography. But it's important to understand that in human geography this de-naturalising move has been seen as morally and politically progressive for the most part. As early as 1974, David Harvey pointed out that those who claim to do things 'for the good of nature' are usually passing their own interests off as if they inhered in the non-human world. Phrases like 'nature knows best' or 'genetic modification is unnatural' all take a supposedly pristine nature as a benchmark against which certain social attitudes or practices are positively or negatively judged. By exposing the social component of both ideas about nature and uses of it, 1990s human geographers were trying to 'de-mystify' collective understandings of the environment (see Box 2.4).

Box 2.4 FEMINIST GEOGRAPHY AND THE ENVIRONMENT

Feminist geographers were among the most important early critics of the idea that the non-human world 'speaks for itself' if only the 'correct' investigative procedures are used to comprehend it. First, several of these geographers showed how both academic and lay understandings of 'natural landscapes' drew upon highly gendered metaphors. For instance, Norwood and Monk's (1987) *The Desert is No Lady* and Kolodny's (1984) *The Land Before Her*, were pathbreaking exposés of the deeply patriarchal assumptions written into dominant views of the US 'frontier' during the eighteenth and nineteenth centuries. In both books, a close scrutiny of these views revealed that frontier lands were seen as things to be tamed, mastered and domesticated to human needs. In other words, dominant male views about women (as the 'weaker sex') were shown to be unconsciously transposed onto views of the natural world – a transposition which further entrenched patriarchy. Second, some feminist geographers broadened this exposure of the pejorative feminisation of the non-human world. For instance, in her important book *Feminism and Geography* Gillian Rose (1993) argued that geography as a discipline is masculinist. Far from geographical knowledge being the result of a disembodied, universal, value-free rationality, she argued that it is highly gendered. This gendered gaze

– in which the geographer is assumed to be objective, unemotional, and clear-sighted – is based, Rose argued, on a set of hierarchical dualisms between culture and nature, mind and body, object and subject, human and non-human that are so taken for granted we forget that they are anything but 'natural'. The masculinism of most geographical knowledge, Rose insisted, was secured by the unwritten assumption that a feminine Other exists that is irrational, emotional, undetached and subjective – an 'unruly' and potentially troublesome Other not dissimilar to the 'wild' frontier landscapes whose depiction Norwood, Monk and Kolodny de-constructed. Rose's book was an important way-station to the now widely accepted idea (in human geography at least) that animates *Nature*: namely, the idea that all knowledge is both constructed and situated. For a full, up-to-date discussion of feminism, feminist geography and the environment see Moeckli and Braun (2001); Rose et al. (1997) is also very informative. These days feminist geographers' research on nature is very heterogenous and difficult to characterise succinctly.

Intellectually, much of the inspiration for human geography's de-naturalising (re)turn to nature came from its engagement with the interdisciplinary field of cultural studies. This field grew prodigiously from the late 1980s onwards and brought together researchers with backgrounds in everything from English literature and communications to philosophy and cultural history. Within this field, three broad frameworks were influential during the late 1980s into the 1990s: namely, postmodernism, post-structuralism and post-colonialism. I cannot do justice to any of these frameworks here so I'll deliberately simplify.[6] Postmodernism, in Jean-François Lyotard's (1984) seminal definition, was a suspicion of 'meta-narratives' and a belief that there are multiple different perspectives on the world not one ostensibly true or correct one (be it science, a religion or what have you). Post-structuralism, closely associated with the works of Roland Barthes, Paul de Man, Jacques Derrida and Michel Foucault, argued that people's subjectivities and their understandings of the world are sculpted by language (rather than simply being expressed by means of language). Sometimes called 'anti-humanist', post-structural thinking located people's identities and beliefs in impersonal 'discursive grids' that varied from society to society over time. Finally, inspired by literary critic Edward Said's *Orientalism* (1978), post-colonial critics argued that colonial power operates not simply through armies, violence or the law but through *representations* of colonial subjects. These representations 'construct' how colonial subjects are seen (and see themselves), which implies that 'de-colonisation' is as much a cultural project as a physical act of the West withdrawing from its former colonies. Together, postmodernism, post-structuralism and post-colonialism drew attention to the *politics of representation*: that is, to who constructs what depictions of the world for what reasons and with what consequences.

These three 'posts', together with a dissatisfaction with Marxist geography and the first wave of feminist geography, allowed a number of oppositional, identity-based branches of human geography to emerge during the 1990s. These included gay and lesbian geography, anti-racist geography, geographies of children and the disabled, subaltern geographies of the non-Western 'Other', and a second-wave feminist geography attuned to the differences among women. Where Marxist and first-wave feminist geography had created a 'social left' – that is, a left-wing human geography concerned with redistributing wealth between social classes and the two genders – the 1990s saw the rise of a 'cultural left' in human geography.[7] This cultural left was concerned with those many groups who are ascribed marginalised or stigmatised identities and, specifically, with how the physical and symbolic content of certain spaces (e.g. the home, the street, the city) reinforced those groups' marginality. Geography's cultural left argued that both power and resistance in society extend well beyond either class or gender. Its rise to prominence in the human side of the discipline can be placed in the context of the so-called 'New Left' in North America and, more generally, the proliferation of 'new social movements' (NSMs) in the West from the mid-1970s onwards. Both the New Left and NSMs were an attempt to broaden the moral and political ambitions of left-wingers away from a rather exclusive focus on (male) workers, class issues and trade union politics (Box 2.5).

What, it may be asked, has all this got to do with nature? In Chapter 1, I mentioned nature's collateral concepts: that is, the other ideas (like 'race') through which ideas about nature find expression. I recall this here because human geography's de-naturalising sensibilities of the 1990s were extended not only to the non-human world but to those things considered

Box 2.5 THE GEOGRAPHICAL LEFT

Since the late 1980s, left-wing human and environmental geographers have become an increasingly important force in geography research and teaching. Though it is difficult to generalise, two things that these geographers have in common are (i) they expose power, domination, inequality, oppression and injustice and (ii) they wish for a future world in which these five things are eliminated or at least ameliorated. One of the peculiarities of geography's left-wingers (so-called 'critical geographers' or 'radical geographers') is that there are few ecocentrists or biocentrists among them. Unlike the disciplines of sociology, philosophy and government/politics, geography's left-wingers rarely preach a 'nature-first' morality – whether in relation to the non-human world or aspects of the human body that might be considered 'untouchable' (like our stem cells). This does not mean that these geographers do not care about what we call 'nature'. Many of them do, but they either (i) seek to balance a concern with the non-human world with a concern for people's well-being (as in the doctrine of 'sustainable development') or (ii) insist that what we call nature does not have *inherent* rights. In the latter case, the argument is that a nature-first morality is a *social choice* that we make, not something dictated to us by the facts of nature (as in moral naturalism, see Box 3.2). For the most part, though, left-wing geographers are not 'greens' (preoccupied with the well-being of the environment) nor are they defenders of the 'natural body' against the 'intrusions' of, say, recombinant DNA technology. Instead, they focus on social, economic, cultural and political issues affecting marginalised or oppressed groups in society – like homosexuals, women or people who suffer from racial discrimination. A graphic illustration of this is Blunt and Wills's (2000) *Dissident Geographies*. This excellent introduction to left-wing thinking in human geography contains no chapter on the environment, while the discussion of identity and corporeality is folded into chapters on constructions of sexuality, class and gender.

by some to be 'human nature' also. For those on the cultural left, the 'human' in human geography was to be understood in thoroughly non-biological, non-essentialist, non-universal ways. This had two dimensions. First, there was an attempt to de-naturalise our understanding of people's identities (sense of self) and their ways of looking at the world. With texts like Peter Jackson's (1987) edited *Race and Racism* leading the way, *Mapping the Subject* ('Thrift and Pile 1995) was just one of several 1990s publications in human geography that showed how people's subjectivity is not explicable in terms of some enduring neurological essence common to all people. Rather, it was shown to be the complex product of the social relations and discourses in which individuals are 'interpellated' (or socialised over the life-course). In this view people unwittingly fit themselves into (and are fitted into) socially created 'subject-positions' over time that are internalised mentally so that they seem to be an organic part of the individuals concerned. Though such arguments may seem to be more the domain of sociologists, social psychologists and cultural theorists, human geographers' concern was with how subjectivities are partly the cause (and effect) of the various physical and symbolic locations in which lives are played out (bars, nightclubs, homes, shops etc.). In particular, 'abject' (or stigmatised) identities were often the focus, like those of gay people or the disabled. The relevance of nature to all this was that discrimination against (as well negative self-understandings among) people with certain identities were shown to often rely upon ideas about what is 'natural' and therefore supposedly 'normal' and what is 'unnatural' and therefore ostensibly 'abnormal'. For instance, until recently, homosexual individuals in the West were led to suppress their sexual preferences (or confine them to certain 'hidden locations') because of socio-cultural conventions that deemed these preferences to be 'perversions' of a universal norm supposedly set by 'human nature' (i.e. that men should be attracted only to women and vice versa).

Second, coincident with these attempts to de-naturalise identity and subjective outlook were attempts by other geographers on the cultural left to de-naturalise the human body. In everyday life, of course, most people tend to think of bodies as biologically fixed and given. Meanwhile, the sciences of the human body – like medicine – have exhaustively analysed the inner workings of the body and the 'outer' faculties of sight, smell, hearing, taste and touch. Such analyses can in turn influence everyday understandings of the body, notably in the form of 'popular' books and documentaries – like the BBC's *The Human Body* series, hosted by eminent medic Sir Robert Winston. As with the issue of identity and subjective outlook, all this may seem a far cry from human geographers' core research and teaching interests. But from the mid-1990s several of these geographers showed how different individuals' and groups' bodily comportment was not a function merely of physiology but also of those same social relations and discourses shaping peoples' subjectivities. These relations and discourses, it was shown, were expressed in and through a variety of sites where people learned how to comport themselves over the life-course (see, for example, Nast and Pile 1998). At a general level, this research on corporeality resonated with some of the humanistic geography research two decades earlier. However, where this early research operated with the rather universal notion of a human body interacting with local environments through smell, touch and taste, the more recent cultural-left research 'de-essentialises' the body and exposes how power relations within society reach into people's biological being not simply their mental self-understanding. In terms of subdisciplinary fields, this kind of research has been primarily conducted by social and cultural geographers, as well as medical geographers (Box 2.6).

Box 2.6 HUMAN GEOGRAPHY AND THE STUDY OF THE BODY

The body, as Elizabeth Grosz (1992: 243) observes, is 'the concrete, material, animate organization of the flesh, organs, nerves, muscles and skeletal structure'. It may seem odd that many human geographers have taken an interest in the human body this past decade or so. After all, one normally thinks that bodies are studied by medics, human biologists and physiotherapists rather than social scientists. But human geographers are not so much interested in the physiology of bodies – for instance, how joints and muscles work or why some people get multiple sclerosis. Rather, they are interested in two other things that directly affect people's bodies. First, one of the main ways in which different societies differentiate people is through selective *representations* of their bodies. For instance, it is no accident that in predominantly white, Western countries black men are often stereotyped as being muscular and over-sexed. Rather than seeing representations of bodies as accurate, human geographers have asked why certain aspects of certain people's bodies gain social salience and come to be valued in positive or negative ways (see, for example, Jackson 1994). Second, representations of bodies profoundly affect how individuals comport themselves. In turn, these modes of comportment can serve to confirm the representations that engendered them in the first place! As the feminist theorist Iris Marion Young (1990) showed in her germinal essay 'Throwing like a girl', people learn to *use* their bodies in ways that often conform with the societal representations into which their bodies are fitted. Geographers like Gill Valentine have investigated how there's a geography to this disciplining of bodily conduct, as individuals learn which modes of comportment are appropriate to which spatial settings. Overall, contemporary geographers interested in the body take issue with the assumption that differences between people are mostly determined by biology. They argue that there is no given 'natural' body that automatically distinguishes people but only pliable *bodies* that vary because of representations and practices that vary over time and space. In this sense, contemporary geographers of the body eschew the abstract universalism of the humanistic geography influential in the 1970s and early 1980s. Most recently, geographical researchers on the body have challenged the distinction between the social and physical dimensions of the body (see Chapter 5). Good summaries of geographical understandings of the body can be found in Valentine (2001: ch. 2), Hubbard et al. (2002: ch. 4) and Duncan et al. (2004: ch. 19). For a wider introduction to social science research into the body see Shilling (1997; 2003).

This de-naturalising research into human identity and the human body can best be understood within the wider context of a 'nature versus nurture' debate in the West that goes back to at least the 1970s. Within the disciplines of human biology, physical anthropology, neuro-psychology and the young field of socio-biology, a debate has raged over whether people's mental and physical capacities are mainly a function of genes and the like or a product of their socio-cultural environment. As this debate has unfolded, new biotechnologies have been invented that promise to alter 'human nature' so that behavioural 'disorders' or congenital diseases can, potentially, be engineered out of existence. In this context, many people worry that we're witnessing a new biological determinism to rival the eugenic beliefs popular in Western countries in the 1920s and 1930s. The concern is that beliefs about the supposed links between a person's genes and their behaviour or appearance will be used to target those with an 'inferior' or 'abnormal' genetic constitution. Clearly, human geographers' recent insights into the mind and body pose a challenge to such deterministic ideas. They show that claims about 'natural kinds' often conceal the biases and the group interests of those who make these claims.

And the rest of geography during the 1990s?

While human geographers (re)discovered nature in a paradoxical (i.e. de-naturalising) way, physical geography during the 1990s remained firmly focused on the natural environment and retained the aspiration to produce 'scientific' (i.e. truthful and objective) knowledge of the non-human world (see, for example, Rhoads and Thorn 1996). Though many human geographers still classified themselves as scientific researchers, it's fair to say that physical geographers used the appellation more widely and unself-consciously than their counterparts did. Previous tendencies towards specialisation in their 'half' of the discipline intensified, in part because new measurement and monitoring techniques produced greater volumes of seemingly more accurate information about particular facets of the environment. Indeed, an ostensibly 'new' branch of physical geography gained momentum through the 1990s: namely, Quaternary studies (i.e. the study of environmental change during the Quaternary era, a recent period geologically speaking). With specialisation came fragmentation, leading some (e.g. Slaymaker and Spencer 1998; Gregory et al. 2002) to call for a more unified physical geography that could trace the interactions between the different environmental 'spheres' (litho-, hydro- etc). The trend toward case-study research also continued (especially in geomorphology), while increased computing-power permitted more complex analysis of ever-larger data sets.

In terms of change (rather than continuity), 1990s physical geography altered in three main ways. First, the balance between pure and applied research arguably tilted slightly in the latter's favour. According to Gardner (1996), this reflected a growth in the 'environment industry' after the Earth Summit and, particularly, the field of environmental management. Issues such as desertification, water pollution, soil erosion and de-forestation increasingly made it onto physical geographers' research and teaching agendas (see Gregory 2000: ch. 7). These geographers often sought to aid environmental managers by pinpointing the physical changes caused by certain human actions (e.g. Burt et al. 1993). Second, at a more philosophical level, physical geography moved away from the 'steady-state' and 'dynamic-equilibrium' assumptions that had underpinned much 1970s and 1980s research. Instead, physical geographers began to appreciate that the environment is complex, often disorderly, and even chaotic in its operations. As Barbara Kennedy (1979) presciently noted in the late 1970s, physical geographers are confronted with a 'naughty world' (see also Kennedy 1994). This change in ontological assumptions was partly inspired by wider shifts in scientific thinking, notably the rise of trans-cendental realism (discussed earlier), as well as complexity and chaos theory (Phillips 1999). Third, the rise of Quaternary studies and a new emphasis on 'global environmental change' meant that the study of environmental systems at large spatial and/or temporal scales underwent a revival. In a sense, physical geography's Davisian origins as what Simpson (1963) called a 'historical science' were rediscovered, providing a counter-balance to the small-scale, process-form studies that had been so popular from the late 1950s. This meant that its credentials as an idiographic subject were reasserted, not at the expense of a nomothetic approach but as a recognition that general laws and processes can have non-general (unique) outcomes (as critical realists argue).

Meanwhile, the human–environment tradition of research continued to be divided between a 'managerialist' and a more radical arm. The latter was represented by TWPE and post-Hewitt hazards analysis, both of which continued to focus on the human dimensions of the human–environment relationship. TWPE moved into a 'second phase' (Peet and Watts 1996) wherein research focused even more on the social and cultural aspects of human usage of the environment (see Braun 2004: 159–63). These two fields of radical human–environment research were also complemented by the rise of five others. First, there were 'environmental injustice' studies, which examined how and why marginal social groups suffer a disproportionate burden of pollution or noxious facilities (e.g.

incinerators) compared to wealthier or more influential social groups (e.g. Pulido 1996). These injustice studies focused on what Beck (1992) called 'manufactured environmental risks', not ones that are wholly natural, and were especially evident in US geography. Second, in Britain, at around the same time, several geographers became interested in how environmental experts communicate their findings to a wider public and how, in turn, the public can democratically inform environmental-policy formulation (e.g. Eden 1996). This research into expert and lay knowledges of the environment was an attempt to challenge the post-war 'linear model' which presumed that scientists and policy-makers know best, with the public positioned as a mere recipient of policies fashioned on its behalf but without its active input. This research showed that environmental knowledges are plural and often conflicting, and it impinged on the issue of 'environmental citizenship' among ordinary people (e.g. Burgess et al. 1998). Third, rural and agricultural geography became radicalised in the 1990s and also ceased to be the intellectual backwaters they had been since the Second World War. Agricultural geography, in particular, became energised by the application of radical ideas from economic geography to the analysis of changes in modern farming – like the move to factory farming of certain livestock in industrialised societies (see Goodman and Watts 1997). Fourth, several geographers developed critical perspectives on how national and local states regulate societal uses of the environment (e.g. Bridge 2000). This research into environmental regulation and governance treated the state as a non-neutral actor interposing itself between business, the public and the natural environment. Finally, some (mostly human) geographers called for an 'animal geography' that would examine the changing character and ethics of people–animal relations over time and across space (Wolch and Emel 1998).

In contrast to all of the above, a less politically radical tradition of resource geography continued to operate through the 1990s, one that can be traced back to the aforementioned early 1970s scares about resource exhaustion. This kind of resource analysis was concerned with how best to use resources given their often finite nature, competing demands for their use, and uneven access to them within and between societies (Rees 1990). Politically, it tended not to ask fundamental questions about the social creation of resource scarcity, the social restriction of access to resources or the widespread belief that resources are but means to human ends (Emel and Peet 1989). This said, a few resource analysts in geography – no doubt influenced by the debates on 'sustainable development' – did take a 'dry green' (or weak ecocentric) perspective on resources and asked for a more drastic alteration of how societies use their natural resource base (e.g. Adams 1996). Similarly, a number of human geographers became interested in 'ecological modernisation', which is the idea that advanced industrial societies can, with proper governmental intervention and a shift in societal attitudes, combine continued economic growth with prudent management of the natural-resource base (e.g. Gibbs 2000). Figure 2.2 and Table 2.1 summarise how contemporary geographers study nature.

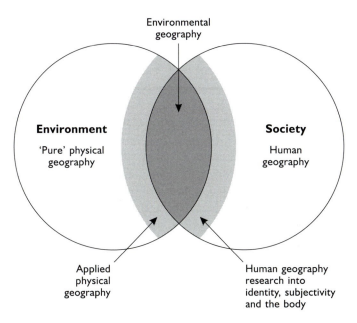

Figure 2.2 Contemporary geography and the study of nature

And today . . .?

As the twentieth century has given way to the twenty-first, there are two notable things about geography's approach to the subject of nature. First, it's clear that human and physical geography produce very different kinds of knowledge about nature – or, rather, about the very different things designated by that word. Physical geography's broadly realist[8] approach remains focused on the environment, as it has been for decades – be it the

Table 2.1 The study of 'nature' within and between geography's three main branches

Physical geography	Environmental geography		Human geography
← Pure research · · · · · · → Applied research	**Status quo/reformist** ← · · · · · → **Radical/left-wing**		**Main subdisciplines** • cultural and social geography • medical geography
	↑ Anthropocentric • resource geography • hazards geography	• radical hazards geography	
		• rural geography • agricultural geography	
• geomorphology • hydrology • biogeography (with soils) • climatic geography • Quarternary geography	• ecological modernisation research • sustainability research	• TWPE • environmental regulation and governance research • expert + lay environmental knowledges research • environmental injustice research	**Main approaches** • second-wave feminist geography • gay and lesbian geography • anti-racist geography • subaltern geography • children's geography • geography of the disabled
	↓ Ecocentric		
The non-human world/ physical environment	*Ideas about, understandings of, and uses of the physical environment*		*'Human nature'*
			Identity/subjectivity ¦ *The body*

humanly altered environment or unaltered environments. Human geography, by contrast, contains many who are suspicious of the idea that the 'facts' of those things we call nature can ultimately speak for themselves. It also contains those interested not in the 'natural environment' but in the 'unnatural environment' created by particular societies at the level of both imagination and reality. And, finally, it contains many who de-naturalise understandings of what is sometimes thought of as 'human nature' (i.e. capacities of the body and mind). Overall, human geographers take a de-naturalising – even *anti*-naturalist – approach to their subject matter. What this means is that, over a century after the 'geographical experiment' was initiated, there is no one overarching theory in geography that explains the relationships between the social and the natural worlds. Instead, we have two different communities of researchers operating with very different theories, models and methods and arriving at very different conclusions about 'the nature of nature'. Sandwiched between these communities is a smaller, rather diverse cluster of human–environment researchers.

For many in the discipline this is nothing short of a tragedy. For them, we live in an era when human and physical geography can fruitfully reunite. The middle ground currently occupied by a smallish number of geographers could, in the eyes of these commentators, become more heavily populated if geographers focused on how proliferating local and global environmental problems can be ameliorated (Cooke 1992). Indeed, on the eve of the Earth Summit, Billy Turner et al. published *The Earth as Transformed by Human Action* (1990) – a sort of agenda-setting follow-up to Thomas's (1956) volume edited out of Clark University (where George Perkins Marsh had plied his trade). In addition, if Gregory (2000: pt IV) is right, physical geography is likely to continue the trend towards applied research. Turner (2002: 61) has thus recently asked: 'Is the discipline poised for another moment in which the human–environment identity ascends to dominance?' (see also Liverman 1999). For several reasons the answer is probably 'no'. Internally, geography remains too diverse and fragmented for a wholesale return to the 'bridging' function envisaged by Mackinder and Davis. Externally, I suspect Gardner's right that geographers 'have been left standing at the bus stop as the ecologists, earth scientists and environmental scientists have rushed to board [the "human impact" bus]' (Gardner 1996: 32). I will say more about geography's human–physical divide in Chapter 4. For now, though, it's worth noting that at the teaching level geography's 'human–environment' identity is much stronger than it is at the research level. Pre-university students are often drawn to geography because of its perceived role as the study of humanly caused environmental problems, and many university teachers and textbook writers cater to this student audience (see, for example, Middleton 1995 and Pickering and Owen 1997).

All this said, in the past five years or so, a cohort of geographers have called into question the ontological distinction between the 'natural' and the 'social' domains. This distinction, as explained earlier in the chapter, is fundamental to the differences between human and physical geography. It is a distinction between *linked*, but putatively *different*, orders of reality. Recently, some have suggested that reality is not, after all, separated into these two ontological domains. They claim that we have always lived in a 'post-natural', 'post-social' and even 'post-human' world. This claim, in rather different ways, has been made by (mostly human) geographers enamoured with actor–network thinking, non-representational theory, process dialectics and the so-called 'new ecology'. I will examine these non-dualistic approaches in this book's penultimate chapter. These approaches have, as we will see, been applied to understandings of the human body and the non-human world. They challenge the ontological division that holds human and physical geographers apart but do not, I shall argue, do much to unify geographers' understandings of nature. This, I will further argue, is a good thing.

SUMMARY: GEOGRAPHY'S NATURES

This chapter has been a very long one. So what have we learnt from this potted history of geography's engagement with the subject of nature? Most obviously we have discovered five things. First, it's evident that geographers' understandings of nature have altered drastically since geography was founded as a university subject. This alteration can only be understood as the combined effect of changes external to the discipline and debates internal to it. Second, it's equally evident that geographers' understandings of different aspects of nature have become more specialised and diverse over time. Third, it's clear that nature has always been something of a problem for geography, both in terms of the discipline's internal constitution and its external links with other disciplines and outside bodies (e.g. funding agencies, user-groups). This chapter has shown that disagreements over which 'nature' geographers should study (and how) have been integral to the successive reconstitution of the 'nature of geography'. Fourth, we've learnt that geography contains little ecocentric thinking, let alone thinking that seeks to 'defend' a putative 'human nature' from the 'predations' of biotechnology and the like. This contrasts starkly with the widespread public sympathy for the well-being of the non-human world, and with those many groups that object to any attempt to alter the 'natural' qualities of people's minds and bodies. Finally, we've learnt that today geography has returned to its original interest in both the environment and 'human nature', yet in ways that are far removed from the era of Davis, Herbertson and Mackinder. Presently, there is no unifying conception of nature that can bring human, physical and environmental geography closer together.

Less obviously, this chapter has, I hope, dispelled two fallacies that the unwary may hold about the knowledges of nature that geographers produce. The first fallacy is to assume that this knowledge has become more accurate and truthful over time. The second fallacy is to assume that while human geographers produce knowledge of other peoples' understandings of nature or the social processes transforming nature, physical geographers produce knowledge of nature 'as it really is' (with environmental geographers producing or combining both kinds of knowledge). Both fallacies privilege physical geography's 'scientific' procedures as the surest way to know nature correctly. They assume that where geographers might have misapprehended nature in the past (in the environmental-determinist days, say) they have got things more or less right today. There are many reasons why these fallacies are just that – fallacies – but two loom large. First, history shows us that what is taken to be truthful or acceptable knowledge about nature in the present may, in the future, be seen as profoundly flawed. Second, since the early 1970s, a band of researchers called 'sociologists of scientific knowledge' (SSK) have shown that scientists' research is never as objective or value-free as it appears. These SSK researchers treat scientists rather as anthropologists treat a foreign 'tribe'. They have shown, often in great empirical detail, that scientists 'construct' their knowledges of nature (unwittingly for the most part) through the philosophical assumptions, theoretical choices and methodological decisions they make. In geography, David Demeritt (1996) has applied SSK thinking to physical geographers' research and, in so doing, arguably reveals the constructedness of all knowledges of nature (see Chapter 4 for more on SSK).

In the remainder of this book I want to look at the main ways nature has been studied by geographers in the past few years. In keeping with arguments made in Chapter 1, I want to focus on the *knowledges of nature* that different geographers produce – not on the 'realities' of the nature these knowledges describe, explain or evaluate. What claims about nature are made in these knowledges? What counts as 'nature' in these knowledges? What are the moral, aesthetic or practical consequences of these different knowledge-claims? What agendas are served when students, professional geographers and other groups in society come to believe some or all of these knowledges are 'correct', 'valid' or 'true'? In posing these questions I particularly want to challenge student readers to take geography – and, by implication, other academic disciplines – off any pedestal it might occupy in their minds. Geography, it seems to me, is one of several domains of knowledge-production that competes with other domains to persuade various audiences that nature is *this* not *that*, that it behaves in *this way*, not *that way*, or that it's moral and/or aesthetic standing is X, not Y and Z. This was my argument in Chapter 1. In none of what follows do I want to deny that there are real things irreducible to knowledge: things that we describe using the label 'nature', or one of its collateral terms. My point, though, is that our comprehension of the 'reality' of those things is deeply conditioned by the sorts of knowledges about them we imbibe, digest and come to accept as legitimate knowledge. In light of this, we need to recognise not only that there's a politics to knowledge (i.e. it's rarely value-free), but that this knowledge also possesses a materiality that is as real as the physical phenomena we happen to label as 'natural' ones. Like sticks and stones knowledge can, indeed, break bones.

EXERCISES

- Compare the contents of the *Dictionary of Human Geography* (Johnston et al. 2000) and the *Dictionary of Physical Geography* (Thomas and Goudie 2000). To what extent is 'nature' a topic of interest in the former and in what ways? Likewise, you might flick through the content of a leading geography journal – like the *Annals of the Association of American Geographers*. Compare an issue from, say, the 1940s with a more recent one. If you look at the essay titles and abstracts, can you identify differences in the kinds of 'natural' things studied and in the way they're studied?

- List some of the key reasons why the ideas academics hold about the world (be they ideas about nature or any other topic) change over time. It might be useful to have a list of 'external' reasons (concerned with what goes on outside academia), a list of 'internal reasons' (concerned with what goes on in academia in general) and a list of 'disciplinary reasons' (concerned with developments internal to a single subject). To get you started, one external reason might be changing public attitudes, one internal reason might be an innovative theory of potentially wide intellectual importance (like Darwinism or chaos theory), and one disciplinary reason might the perennial efforts of young academics to debunk the wisdom of their elders as they quest after fame within their subject area.

- As this chapter has shown, geography has been selectively influenced by wider intellectual and real-world events when it comes to changing understandings of nature in the discipline. Can you explain why geography has often failed to respond to wider developments that have obvious relevance to the understanding of nature – such as the outpouring of ecocentric sentiment in the early 1970s? What might explain why geography has only been *selectively* influenced by its wider societal context? Is it important for geographers to adjust their present-day research in light of the moral and practical concerns attaching to new developments like biotechnology?

FURTHER READING

Chapter 1 of Livingstone's (1992) The Geographical Tradition offers an exceptionally good discussion of 'contextual histories' and why they are be preferred to 'internalist' ones. Heffernan (2003) offers a brief contextual history of early university geography. Those interested in the evolutionary cast of early geographical thinking should consult chapters 6–8 of Livingstone's book plus the entries on environmental determinism, Darwinism and social Darwinism in The Dictionary of Human Geography (4th edn, Johnston et al. 2000). For thorough histories of human and physical geography respectively see Johnston and Sidaway (2004) Geography and Geographers and Gregory (2000) The Changing Nature of Physical Geography. For a potted history of human geography see Hubbard et al. (2002: chs 2 and 3); for potted histories of physical geography Slaymaker and Spencer (1998: ch. 1), Gardner (1996), Sims (2003) or Inkpen (2004: ch. 2). The postwar debate revolving around Hartshorne and Schaefer is dealt with succinctly by Unwin (1992: ch. 5). Derek Gregory's (1978) Ideology, Science and Human Geography (ch. 1) still offers one of the best general introductions to spatial science, even if it generalises a little too much. Johnston (2003) and Richards (2003a) discuss human and physical geography in relation to the social and natural sciences respectively. The relevant chapters of the following books offer excellent introductions to humanistic, Marxist, feminist and post-geographies: Modern Geographical Thought (Peet 1999); Dissident Geographies (Blunt and Wills 2000) and Approaching Human Geography (Cloke et al. 1991). For readers interested to know more about the thinking of some of the geographers mentioned in this chapter a series of 'biobibliographical studies' have been published annually since 1977. Unfortunately, apart from this chapter, there is no other single source that discusses the history of geographers' engagement with the topic of nature. Though very good, Olwig's (1996) discussion of nature and geography rests upon a limited definition of the term nature. Finally, Beaumont and Philo (2004) and Eden (2003) discuss geography's (non-)engagement with the environmental movement and ecocentrist perspectives.

3

DE-NATURALISATION
Bringing nature 'back in'

'The one thing that is not "natural" is nature'.

(Soper 1995: 7)

INTRODUCTION

In this and the following two chapters I want to explore the nature-knowledges produced by contemporary geographers in some detail. My tack is to remain agnostic about the truthfulness or validity of these knowledges.[1] Instead of taking sides, I stand back and ask how and why different geographers depict 'nature' in the ways they do, with what effects and to what ends. I leave it to readers to judge the relative (de-)merits of the nature-knowledges these geographers are currently producing. Since I lack the space for a comprehensive discussion, I've chosen to focus on the principal ways that nature has been studied in geography over the past decade or so. This parsimonious approach involves identifying key themes rather than surveying the literature in detail. In this chapter I explore what I argued in the previous one is a dominant theme in both human geographers' and many environmental geographers' recent research into nature: namely, its 'de-naturalising' thrust. In Chapter 4, by contrast, I examine the argument of (mostly) physical geographers that nature is a 'real' domain that is not, in either degree or kind, a 'social construction'. After concluding this chapter with an examination of the 'human–physical divide' in geography, the penultimate one looks at the work of those who would do away with the idea that nature is either a social construction or a reality irreducible to social representations and practices. In effect, this work transcends the commonplace dualisms of 'nature'/'society', 'human'/'non-human'. In each chapter, I try to use case studies and vignettes in order to illustrate how geographers are studying nature in the ways they are and why it matters. I draw these examples from the published literature so that readers can see exactly how nature in the 'real world' has been analysed by geographers. My trawl of this literature is deliberately selective since I do not want to sacrifice depth for a broad-brush discussion of the knowledges of nature in question.

As I said, this chapter describes and explains how several contemporary geographers have de-naturalised nature. As we'll see, this involves one or both of the following theses, depending on which geographers are doing the arguing: (i) that nature is less important as a causal factor in human affairs than was previously thought and (ii) that those things that seem to be natural are, in fact, social through and through. In its strongest form, the de-naturalising argument suggests that nature is not natural (i.e. only *apparently* natural). This amounts to a reversal of environmental determinism over a century after geography's constitution as a university subject.[2] It suggests that societies, whether they realise it or not, hold the key to understanding what nature is and what happens to those things we think of as natural things. Geography's de-naturalising thrust has been led by critical human geographers and by a cohort of left-wing environmental geographers. In this light, my chapter title can be seen as an ironic one. For these geographers have 'rediscovered' nature as a topic in a subversive way: namely, by questioning its very naturalness. As we'll see, this questioning has not been absolute. There are several human and environmental geographers who still believe that what we call 'nature' matters in a biophysical sense. But for the most part, present-day human geographers and several environmental geographers hold to a 'nature-sceptical' (de-naturalising) stance.

It is interesting to compare this stance with that held by others in the social sciences and humanities. For the most part, economists have jumped on the 'human-impact' bandwagon. This is no doubt because anthropogenic environmental problems appear to be proliferating and also because a good deal of funding is available for research into how to ameliorate these problems. Environmental economics has emerged as a distinct field since the late 1960s and aims to alter economic practices so as to reduce their environmental impact (see, for example, Bateman and Turner 1994). Similarly, many researchers in the discipline of politics and government have examined how and why political actors deal/fail to deal with environmental problems (e.g. Young 1994). It is more or less accepted that these problems are real and in need of a solution. The disciplines of anthropology, philosophy and sociology have been more ambivalent. On the one side, all three have traditionally focused on the human world, leaving the study of nature (non-human and human) to the physical, materials and medical sciences (though anthropology has a strong 'human–environment relations' focus). Recently, though, all three disciplines have belatedly recognised that some societies appear to be having an unprecedented impact upon the environment. Several anthropologists, philosophers and sociologists have inquired into the beliefs, values and practices that cause environmental degradation, with philosophy containing the most outspoken ecocentric critics (e.g. Light and Rolston 2003). Analogously, some have also inquired into why the boundaries of the human body are apparently being broken down and have scrutinised the propriety of this boundary transgression. Again, philosophers have taken the lead in evaluating whether everything from cloning humans to in-vitro fertilisation is morally acceptable (e.g. Burley and Harris 2002). On the other side, though, like the geographers whose work is discussed in this chapter, several others in the three disciplines argue that societies construct nature at the level of both representation and materiality. For instance, sociologists of the environment and the body have shown how different societies produce different understandings of and effects upon the natural world (e.g. Macnaughten and Urry 1998). Currently, then, the social sciences are divided in their approach to nature (between and within themselves): to use Soper's (1995) terms once more, 'nature-endorsing' positions are opposed by 'nature-sceptical' ones. Things are rather more clear-cut in humanities subjects like English and cultural studies. With the notable exception of environmental history, most of these subjects – when they consider nature at all – tend to focus on social representations, discourses and images of it in different times and places.

If we take geography as a whole it combines both of Soper's positions in one discipline. The nature-endorsing stance holds that (i) there is a real world of natural phenomena, (ii) the properties of that world are knowable in a relatively objective way, and (iii) we can derive (though not necessarily directly) moral (and aesthetic) judgements about nature from an understanding of the 'facts' of this world. Physical geographers, as we'll see in the next chapter, hold fast to (i) and (ii) above, leaving a small number of others in the discipline to focus on (iii). If we're to evaluate the claims of these geographers – who insist that their knowledge of nature is more or less accurate – we need first to understand the nature-sceptical stance of many others in the discipline. I begin by discussing the precedents for the de-naturalising knowledge of nature I explore in this chapter. Thereafter, I use case studies to explore the key claims of geography's present-day nature-sceptical researchers.

PRECEDENTS

In the previous chapter I noted the inverse correlation between wider societal concerns with nature and the level and character of geographers' interest in nature. Since the late 1960s there have been two waves of environmental concern, the latter (beginning in the early 1990s) coincident with increased anxiety about science and technology's new-found power to remake human nature. Yet only parts of geography have embraced the 'human-impact' agenda, while very few geographers seem alarmed by the recomposition of the human person prefigured by molecular genetics, nanotechnology and the like. This is not to say that geographers don't (in their professional capacity) care about what's happening to those things we call natural things. But it does raise the following question: why, when so many people are worried about the possible 'end of nature', do many geographers seem relatively unconcerned? The answer is, I think, twofold. First, as I'll explain in the sections to follow, these geographers doubt whether there is a 'nature' whose end is nigh. Second, these geographers argue that when nature-talk proliferates in a society we should inquire into who is doing the talking and what they have to gain (and lose) from discussing nature in the ways they do. In Chapter 1, I argued that the word nature has three principal meanings. As early as 1974, during the first wave of environmental concern in the West, David Harvey argued that the powerful in any society wield the definition of keywords like nature to their own advantage. His essay 'Population, resources and the ideology of science' was unorthodox in its time and set a precedent for the current attempts by geographers to de-naturalise nature. Where many geographers of this era tried to reinvigorate the disciplinary middle ground by focusing on natural resource exploitation, Harvey argued that the supposed 'environmental crisis' of the early 1970s was a fiction. Somewhat later, Kenneth Hewitt's (1983) *Interpretations of Calamity* sought a less biophysical explanation of what seemed to be a quintessentially natural phenomenon: namely 'environmental hazards'. In this section I want to explore Harvey's and Hewitt's early interventions because they focused attention on two broad issues that preoccupy contemporary researchers in human and environmental geography: namely, the power of representations of nature and, second, the relative causal importance of natural and societal processes (Whatmore 1999).

Ideologies of nature

As the twentieth century gave way to the twenty-first, population history was made. In mid-1999 the world's populace numbered 6 billion for the first time. Shortly thereafter India became the second country on the planet to contain over 1 billion inhabitants. Looking ahead, the United Nations predicts that the global population will number some 9.3 billion by 2050, a 200 per cent increase on the 1950 total. Though one rarely hears the term in official circles these days, such figures lead some to worry about 'overpopulation'. This notion can be traced back to the influential writings of English economist and demographer Thomas Malthus (1798). Malthus maintained that while resources can only be increased in an arithmetical progression (2, 4, 6, 8 etc), population numbers can increase geometrically (2, 4, 8, 16 etc.). In the modern era, overpopulation thinking is associated with the neo-Malthusians of the early 1970s. As mentioned in the previous chapter, alarmist books like *Blueprint for Survival* (Goldsmith et al. 1972), *The Population Bomb* (Ehrlich 1970) and *The Limits to Growth* (Meadows et al. 1972) predicted a dire future where a finite natural-resource base would limit the numbers of people who can live on the planet. Following Malthus's thinking, these books argued that 'preventative checks' (like limiting childbirth through increased contraceptive measures) were the only way to avoid 'positive checks' like starvation. The context for these and other neo-Malthusian analyses was rapid post-1945 population growth in the developing world – notably in Africa and Asia. The stark practical implications of neo-Malthusianism were well captured in a 1974 essay by American biologist Garret Hardin. In 'The ethics of a lifeboat' Hardin allegorically depicted the West's rich population as living on a lifeboat surrounded by the poor of the developing world who are desperately trying to clamber aboard. Since room on the lifeboat is limited, Hardin argues that the only ethical solution is to ignore the poor's pleas for help so that those on the boat can survive. The benevolent alternative – namely, trying to assist the floating masses – would, in Hardin's view, bring ruin to all. As he put it:

> we should go lightly in encouraging rising expectations among the poor . . . for if everyone in the world had the same standard of living as we do, we would increase pollution by a factor of 20 . . . Therefore it is questionable morality to increase the food supply. We should hesitate to make sacrifices locally for the betterment of the rest of the world
>
> (Hardin cited in Neuhaus 1971: 186)

Before turning to Harvey's critique of neo-Malthusianism, we need to make explicit the conception of nature implicit in overpopulation arguments. The Activity below will get you thinking about how nature is conceived in neo-Malthusian reasoning.

ACTIVITY 3.1

Read the previous paragraph again and try to answer the following questions:

- Which of the three definitions of nature stated in Chapter 1 are part of the overpopulation argument?
- How are these definitions linked to moral judgements and practical policies?

Let's answer each question in turn. The term nature, you will recall, means (i) the non-human world, (ii) the essence of something, and (iii) an inherent force ordering the human and non-human worlds. All three definitions are in play in neo-Malthusian reasoning. First, the non-human world of natural resources is cited as a key factor limiting population growth. Second, these resources are seen as both quantitatively and qualitatively finite: it is in their nature (their essential character) to be non-ubiquitous. Third, neo-Malthusianism sees the propensity of people to breed beyond the natural-resource base as a 'natural law' that can only ever be tempered but never fully eliminated. Here nature is seen as creating a dynamic balance between population numbers and resource availability over time and space. On the basis of these claims about what nature is (or how it behaves), neo-Malthusianism draws some direct moral and practical lessons. In other words, it connects *facts* to *values* and what *is* to what *ought to be done*, as if values and actions can be 'read off' from the supposed 'realities of nature'. For instance, Hardin's tough stance on whether or not to help the populous developing world follows *logically* from his belief that the more people there are on the planet the less there is to go around.

During neo-Malthusianism's heyday, there was some evidence to support the overpopulation argument. Escalating birth rates in the developing world were clearly correlated with a rising (or at least not declining) incidence of malnutrition, starvation and famine in many countries. This evidence, in tandem with the simple, intuitively appealing logic of the overpopulation argument, made neo-Malthusianism a real intellectual force in many academic disciplines, in several political parties (for instance, Indian governments sponsored male sterilisation programmes in the 1970s) and in the wider society. In this context, David Harvey's anti-Malthusian reading of the population–resources relationship was strikingly unorthodox. Inspired by the ideas of the radical nineteenth-century economist Karl Marx, Harvey argued that neo-Malthusianism was an *ideology*. According to Marx, the ruling ideas of any era are the ideas of the ruling classes. For Harvey, the reason that neo-Malthusianism became so influential in the early 1970s was not because it was objectively true but, rather, because it served the interests of Western elites to *claim that it was objectively true*. Let me explain.

Harvey acknowledged that within its own terms of reference neo-Malthusianism made sense. It comprised both abstract 'logical truths' (e.g. if resources are assumed to be finite and if population is assumed to grow geometrically then it follows that eventually overpopulation will result) and 'empirical truths' (e.g. facts about population growth rates, malnutrition and mortality in various countries). If the latter seem to correspond to the former – as they did to many in the early 1970s – then it's no surprise that neo-Malthusianism appears to be a plausible explanation of the population–resources relationship as well as a logical policy response. However, Harvey's criticism was not that the logic of neo-Malthusianism was flawed nor even that the evidence supposedly confirming that logic was erroneous. Instead, he took issue with the *assumptions about nature* that underpinned the whole overpopulation argument. First, he questioned the idea that the amount of natural resources people need to subsist is determined by their biological needs. Subsistence levels are, he insisted, defined *relative to* a person's 'historical and cultural circumstances' (1974: 235). Thus the bundle of resources deemed necessary to subsist in one society at one moment in time will be very different to others in the present and future. Second, Harvey argued that 'natural resources' are socially, culturally and economically defined. Certain things only *become* resources when a particular society has the means and the desire to utilise them; until then a naturally occurring phenomena is not a resource for that society. Finally, Harvey argued that resource scarcity is not given in nature but, rather, is the outcome of societal processes. This created scarcity arises, Harvey argued, because of power relations internal to society wherein some social groups command far more wealth than other groups. More specifically, Harvey's Marxist viewpoint suggested that in capitalist societies both the lower cadres of the working class and the unemployed are denied the monetary wealth to purchase the means of subsistence. Thus, what neo-Malthusians called 'overpopulation' was, for Harvey, a 'relative surplus population' produced by capitalism's tendency to create poverty for the many and wealth for the few.

In effect, Harvey argued that unproblematised assumptions about nature were used as a smokescreen to justify the West's unwillingness to redistribute wealth to the developing world. For him, neo-Malthusianism was an ideology in the double sense that (i) it concealed the truth about the population–resources relationship and (ii) it justified a Western elite's determination to concentrate global wealth rather than share it with needy developing-world populations. By licensing population-control policies or else 'benign neglect' (Hardin's preferred option), Harvey saw neo-Malthusianism as a cunning way of justifying the poverty of the poor and attempts to monitor their reproduction. As he put it, 'whenever a theory of over-population seizes hold in a society . . . then the non-elite invariably experience some form of . . . repression' (1974: 237). Harvey's Marxist interpretation of 'overpopulation' was designed to expose the truth that neo-Malthusianism obscured and to offer very different value judgements about, and policy responses to, the phenomena of 'natural resource scarcity'. To quote him at length:

> let us consider a [neo-Malthusian] . . . sentence: "Over-population arises because of the scarcity of resources available for meeting the subsistence needs of the mass of the population". If we substitute our definitions [of subsistence, resources and scarcity] into this sentence we get: "There are too many people in the world because the particular ends we have in view (together with the form of social organization we have) and the materials available in nature, that we have the will and the way to use, are not sufficient to provide us with those things to which we are accustomed". Out of such a sentence all kinds of possibilities can be extracted:
>
> 1. we can change the ends we have in mind and alter the social organization of scarcity;
> 2. we can change our technical and social appraisals of nature;
> 3. we can change our views concerning the things to which we are accustomed;
> 4. we can seek to alter our numbers
>
> . . . To say that there are too many people in the world amounts to saying that we have not the imagination, will or ability to do anything about propositions 1, 2 and 3
>
> (1974: 236)

Harvey's critique of neo-Malthusianism was among the first in geography to show that ideas about nature are not innocent in relation to the world they purport to describe, explain and evaluate. His notion of ideology – that is, a set of ideas that appear to be true but which in fact conceal the truth in order to further a certain groups' interests (see Box 3.1) – paved the way for later research into how ideas of nature do not reflect the realities of nature but, rather, the societal contexts in which those ideas arise. It's important to note that Harvey was not denying the physical existence of those things we call natural (in his case, resources). After all, Marx, Harvey's chief inspiration, was a self-proclaimed 'materialist' who believed that a real world exists regardless of our ideas about it. But these ideas matter because it is precisely through them that we come to understand that material world. After Harvey's intervention, the next major attempt to 'de-naturalise' representations of the non-human world prior to the current period was, arguably, that of Denis Cosgrove and Stephen Daniels (Box 3.2).

Box 3.1 IDEOLOGIES OF NATURE

The term 'ideology' has multiple meanings. Broadly speaking, it is used pejoratively by left-wing intellectuals, notably Marxists. This wasn't always so. The term dates from late-eighteenth-century France where it meant simply the study of ideas and, more specifically, ideas that were free of religious or metaphysical bias. However, because of the influence of Marx and his co-author Friedrich Engels, the term 'ideology' took on a more particular and negative meaning. Some Marxists saw ideologies as distorted systems of ideas (leading to 'false consciousness' among those who believed these ideas) that are promulgated by powerful social groups in order to deceive the mass of society as to their 'true interests'. In Harvey's critique of neo-Malthusianism there is a suggestion that this understanding of ideology is in play. After all, he implies that his Marxist interpretation of the population–resources relationship is somehow 'better' (more accurate?) than the neo-Malthusian one. However, he qualifies the 'false consciousness' idea by declaring that one can never step outside ideology in order to inspect nature 'as it really is'. This suggests a broader conception of ideology defined as any set of ideas designed to facilitate certain social interests by depicting the world in a selective way. This broader conception arguably animates the 1984 work of Harvey's student Neil Smith. Smith's *Uneven Development* (1984) makes formal reference to 'ideologies of nature'. For him, these are 'common-sense' beliefs about nature whose partiality and bias is dissimulated precisely because they seem to have no social contamination – because they seem to be about nature in itself not society. In recent years, left-wing analysts of all stripes have used the term 'ideology' in an ever looser, less precise way. At the same time, the term now pops up in all manner of discussions across the political and moral spectrum. These days, among left-wing analysts at least, it is routinely interchanged with the terms 'hegemony' and 'discourse'. I'll say more about these two terms in the next section of this chapter.

Box 3.2 THE DE-NATURALISATION OF LANDSCAPE

Around a decade after Harvey's essay (1974), Denis Cosgrove put forward an arresting argument about one of nature's collateral concepts and one of geography's main objects of analysis: namely, landscape. In *Social Formation and Symbolic Landscape* (1984), he argued that landscapes are not simply physical environments existing 'out there' for people to see, study, use or enjoy. Instead, he argued that landscape is a specific 'way of seeing' coincident with the emergence of capitalism in Europe from the sixteenth century onwards. When we think of the word 'landscape' we often think of fields, water courses, trees, sky, fields and livestock arrayed before us. Cosgrove argued that we have, historically, learnt to see the apparently objective facts of landscapes in a certain way. From the period of the European Renaissance, capitalism began to supplant previous modes of production, while the invention of three-dimensional perspective and new cartographic and surveying techniques permitted a new way of representing urban and rural spaces that fast became 'common sense'. Cosgrove showed how newly wealthy urban merchants and industrialists purchased estates in the countryside and began to commission paintings of their properties. These paintings typically contained little or no human presence, gave the viewer a detached all-seeing perspective on a 'natural' panorama, and appeared to be highly realistic. Cosgrove's point was that the view of landscape here was both constructed and highly particular. For him, it not only reflected the landowner's desire to match his physical ownership with visual ownership. It also deliberately made invisible the work of peasants and rural labourers who were often dispossessed so that urban elites could enjoy their picturesque views of a seemingly harmonious, well-ordered rural environment. By 'naturalising' the view, landscape painting thus, for Cosgrove, both arose from and reproduced the social relationships of a nascent, class-divided capitalist society. In his estimation, landscape was a class-specific way of seeing akin to 'ideology' in the Marxist sense of the term. Along with Stephen Daniels, Cosgrove went on to pioneer the geographical study of 'symbolic' and 'iconographic' urban and rural landscapes. This research opened the door for the 'culture of nature' thinking I discuss later in this chapter. It also pointed to the importance of *visual* constructions of nature, as much as written and spoken ones.

Unnatural hazards: de-emphasising the physical environment

Aside from showing that ideas about nature are constructed (rather than being 'mirrors of nature'), Harvey's essay also questioned the relative causal importance of the environment in understanding human–environment relationships. In other words, once one had penetrated behind the veils of ideology, Harvey argued that the environment is not as important a factor as is often supposed in the environment–society relationship. Specifically, his critique of neo-Malthusianism implied that *what appear to be naturally caused problems (like starvation) are, in fact, socially caused problems*. This attempt to de-emphasise the physical environment was central to *Interpretations of Calamity* (Hewitt 1983). The importance of this book is that it sought to de-naturalise 'natural hazards' – which, by definition, appear to be thoroughly non-human and non-social in their origins. What, it might be asked, can be non-natural about droughts or tornadoes or floods? Or rather: why do we routinely think that natural hazards are just that, 'natural'?

If we take earthquakes as our example, then an answer to the Activity question might be as follows. First, earthquakes are natural because they are caused by geophysical processes that humans can do little to influence. These processes unfold well below the earth's surface where continental plates torque and collide. Second, earthquakes occur regardless of what people think about them and regardless of whether or not people experience them. In short, earthquakes seem self-evidently natural and they become 'hazards' if and when people suffer their effects (like collapsed buildings). A recent, shocking example that appears to confirm this was the earthquake that hit the ancient city of Bam in Iran. Measuring almost 7 on the Richter scale, the December 2003 earthquake killed a staggering 40,000 people and injured thousands of others.

This discussion of the naturalness of 'natural hazards' may seem obvious and uncontroversial. If we took any number of natural hazards – floods, hurricanes or tsunamis, say – most people would readily agree that they are primarily natural events. Within the world of hazard analysis and management, what Hewitt called the 'dominant view' comprised the following beliefs (1983: 5–9):

- Hazards are extreme natural events, low in frequency but high in magnitude.
- Because the impact of hazards can be ameliorated but rarely controlled, hazards are independent variables to which societies must adapt and adjust.
- Hazards are best managed using technical means that either stymie the geophysical causes of those hazards or else reduce the physical impacts of hazards.

In sum, Hewitt saw the dominant view as fixated on technical solutions to what were seen as naturally occurring events that were largely unpredictable and capricious. In this view, the risk people ran of being harmed by natural hazards was determined by the hazards themselves and, secondarily, by protective measures implemented by individuals and communities.

Plausible though the dominant view seems at first sight, Hewitt (1983: 29) argued that it was the 'single greatest impediment to improvement in [the] . . . quality and effectiveness' of hazards analysis and hazard management. We can understand Hewitt's assertion by looking at the findings of Michael Watts (of the University of California, Berkeley), one of the contributors to Interpretations of Calamity. His illuminating chapter in the book seeks to answer the following knotty question: why do societies that have, in the past, successfully adjusted to certain 'natural hazards' suddenly find themselves vulnerable to the effects of these hazards? One possible answer to this question is that the hazards have become more extreme (i.e. greater in magnitude). Another is that the societies in question have somehow lost the expertise and knowledge to deal with the hazards in question. In Watts's case study – focused on Hausa peasants in northern Nigeria during the nineteenth and twentieth centuries – neither situation applied. What, then, could possibly explain the Hausa's increased vulnerability to drought (the particular hazard Watts focused on)?

To answer this question Watts looked at events internal to society not those pertaining to the physical environment. Like Harvey, he drew on Marxist ideas and applied the concepts of mode of production and moral economy to his analysis of the Hausa. A mode of production is the specific way in which a society organises its productive activities and comprises productive classes (those engaged in producing goods), relations of productions (the specific relations between productive classes), means of production (the principal technologies used in production) and production goals (the ends that production serves). A moral economy comprises the norms, beliefs and values that lend order and coherence to the relations of production and hence to the mode of production as a whole. In pre-colonial Nigeria, the mode of production was a peasant-pastoral one, based on the cultivation of sorghum and millet. Households produced crops for their own subsistence needs, but gave part of their surplus (or else their labour) to village heads who, in turn, were answerable to district heads and thence up to thirty or so emirs who governed the Sokoto caliphate – a Muslim confederation with its own laws, customs and armies. Within this network of vertical and horizontal production relations, crops were produced mainly for their use-value (i.e. for direct consumption within the caliphate) using basic implements in a labour-intensive way.

Northern Nigeria was (and remains) semi-arid and 'extreme climatic variability, particularly drought, is and was an intrinsic part of nature [in this area]' (Watts 1983: 247). This being so, how were peasant households able to survive drought periods when the hierarchical mode of production in which they were embedded 'creamed off' a portion of their crops annually? Here Watts emphasised the importance of the Hausa's moral economy. While this moral order required tribute from households to their overlords, at times of extreme climatic stress a norm of reciprocity was activated wherein emirs, district and village heads would redistribute stored food back down to the household level as and when necessary. In this way, Hausa society created a buffer that ameliorated the impacts of drought.

All this changed subsequent to British colonisation of Nigeria from the early twentieth century. The Hausa experienced major famines in 1914, 1927, 1942 and 1951 – whereas they'd experienced virtually none the century before. While rainfall variability was no more (or less) extreme than in previous decades, Watts argued that the imposition of a capitalist mode of production on the Hausa – achieved through colonial domination – made households far more vulnerable to the effects of drought. In brief, a capitalist mode of production is geared to the sale of commodities for money with a view to making profit. It involves relations between those who own the means of production and those who work for them for money. In addition to this 'primary' class relationship, there are 'secondary ones', also mediated by money (like those between landlords and tenants, or money-lenders and borrowers). Colonialism, whose heyday has now passed, involved the formal occupation of one territory by the government of another or its representatives. According to Watts, the capitalism–colonialism nexus transformed Hausa society in four main ways. First, the colonial authorities promoted the cultivation of groundnuts and cotton among peasant households, replacing the subsistence crops of sorghum and millet. Second, these crops were grown for export to Britain and elsewhere. Third, exchange in kind was supplanted by exchange for money, as Hausa crops entered a cash economy extending well beyond Nigeria. Finally, in 1910 the British imposed a tax on households to be paid in cash not in crops or labour.

Together, Watts shows that these four changes conspired to remove the drought buffer present in pre-capitalist, pre-colonial Hausaland. First, as households switched to cotton and groundnut production they lost control of their traditional food sources. Second, now reliant on earning enough money to buy food and pay colonial taxes, households found themselves subject to the vagaries of international commodity markets. If groundnut and cotton prices fluctuated then Hausa peasants could find themselves lacking the monetary means to buy enough food. Finally, the previous moral economy was eroded as former village and district heads used their wealth to become money-lenders to peasants in need of loans to tide them over. As time passed, former relationships of reciprocity were thus superceded by commercial relationships where loan repayments were expected with interest. As Watts concluded, the Hausa became more vulnerable to drought not because this 'natural hazard' was unavoidable but because of changes in the constitution of the real and moral economies. As with Harvey, Watts wasn't denying the reality of drought (or any other natural hazard). Rather, he saw it as a 'trigger' for problems that were fundamentally socio-economic and political in origin rather than environmental (see Abramovitz 2001; Pelling 2001).

RE-PRESENTING NATURE

I have dwelt upon the ideas of Harvey, Hewitt and Watts at some length because they were insightful precursors to the present-day research by geographers that aims to de-naturalise that which seems natural. In this section I want to focus on claims that what we call nature is nothing more than a set of ideas or representations. In the next section I will focus more on the 'real nature' that these ideas and representations denote. As will become clear, my overall argument in this book that knowledges of nature are not reducible to the material things they refer to resonates with the work of several authors I discuss in the three subsections below. But this does not mean that I take sides and uncritically endorse these authors' ideas. Though I am obviously sympathetic to the notion that what we call 'nature' (either directly or by way of the term's collateral concepts) says as much about our ways of thinking as it does about nature itself, it would be inconsistent of me to champion those who have promulgated this notion. After all, the claim that we often confuse representations of nature with their referents is itself a knowledge-claim: a claim about other people's claims about nature. Accordingly, it is incumbent upon me to remain as impartial as I can be. Among other things, this involves an honest look at the intellectual, moral and aesthetic agenda that those (like me) who insist that nature is a re-presentation are trying to further by writing books like this one.

In contemporary geography there are, broadly speaking, three main variations on the idea that our conceptions of nature are just that: conceptions that we routinely confuse with the things they denote. First, some have focused on 'myths' and 'orthodoxies': that is, false beliefs that nonetheless become influential. Second, other geographers have shown how ideas of nature are woven into the process of hegemony: that is, rule by consent rather than coercion (see Box 1.4 again). Finally, still other geographers maintain that what we call nature is an effect of discourse, wherein representation and reality 'implode'. I want to preface my discussion of this trio with a brief comment on representation. There are many ways in which we re-present nature to both ourselves and others. There is speech, there is writing, there is imagery and there is also sound. In society, people convey understandings of nature through everything from song and poetry to film and novels. When it comes to those things we classify as natural, these various forms of representation all arguably have two things in common. First, because nature cannot speak for itself – be it our bodies, a dolphin, a tree or microbe – we must *speak for it*. In other words, we routinely re-present nature in the sense of being its representative, just as a politician stands for his or her constituents. Second, any act of speaking for those things said to be natural inevitably involves a second element of representation: a *speaking of*. This entails depicting, framing or staging nature in ways that the person doing the representing thinks is most fitting. For instance, where a marine biologist might represent a minke whale in purely cognitive and factual terms, an Earth First! activist might prefer a morally charged depiction of the whale's dignity, beauty and majesty. In sum, what literary critics call the 'double session' of representation involves nature's representers serving as *both* proxies (representatives of it) and stage-managers (selectively depicting nature's 'actual character' – see Woods [1998] for an example of this double session in action). With these two points about the representation of nature in mind, let us now turn to the three ways nature-representations have been understood in contemporary human and environmental geography.

Truth, falsity and nature

In the previous section I presented Harvey's thesis that ideas about nature are often ideological. In Box 3.1 I discussed the notion of ideology and noted that for some analysts it connotes false or deceptive beliefs about the world. In a recent essay explaining what 'the social construction of nature' means, the environmental geographer David Demeritt (2002) identifies two kinds of 'construction talk' in contemporary human and environmental geography. The first he calls 'construction-as-refutation' (I'll come to the second later in the chapter). The geographers who talk about nature in this first way seek to expose erroneous and misleading beliefs about the 'nature of nature'. In this sense, these geographers continue the tradition of ideology criticism inaugurated in Harvey's essay, even if they rarely use the term ideology themselves. For these critics (who, like Harvey, are often left-wingers), nature is 'constructed' not so much physically as at the level of representation. For them representations condition how we understand the nature of nature and, in this sense, even erroneous representations are influential if they go unchallenged for long enough. Therefore, when these geographers show that certain representations of nature are simply wrong, they are refuting them by exposing the social bias distorting their accuracy. In this context, then, the term 'construction' refers to the way that knowledge of nature is manufactured by certain people rather than being a passive reflection of reality.

Good examples of this exposure of false representations of nature are not hard to find in geography. In environmental geography, Third World political ecologists have done much to debunk what they term 'environmental myths' and 'environmental orthodoxies'. According to Tim Forsyth (2003: 38), of the London School of Economics, these myths and orthodoxies are 'generalized statements . . . [about] environmental degradation or the causes of environmental change that are often accepted as fact but which have been shown by field research to be biophysically inaccurate . . . [while] leading to [misguided] environmental policies'. Desertification, de-forestation and soil erosion are just three well-known 'environmental problems' in the developing world that, according to Forsyth and others, have been profoundly misunderstood. This raises two questions: first, why do environmental myths and orthodoxies catch on?; and second, how can their inaccuracy be exposed and environmental policies based on them accordingly dismantled? Detailed answers have been provided in the excellent books *Misreading the African Landscape* (Fairhead and Leach 1996), *The Lie of the Land* (Leach and Mearns 1996), *Desertification: Exploding the Myth* (Thomas and Middleton 1994), *Uncertainty on a Himalayan Scale* (Thompson et al. 1986) and *Critical Political Ecology* (Forsyth 2003; see also the essay by Bassett and Zueli 2000). Here I simply use a case study to tease out some of the key issues.

Northern Thailand lies on the eastern extremity of the Himalayan mountain range (see Map 3.1). It comprises a series of lowland areas in which irrigated rice has been grown for centuries. These areas are surrounded by forested uplands that have been cleared for agricultural purposes by the Karen (an ethnic group indigenous to the uplands) and by migrants from neighbouring China, Laos and Myanmar. For over twenty years, environmental policy in northern Thailand has been influenced by the 'theory of Himalayan environmental degradation'. According to this theory, high rates of population increase in the Himalayan region routinely lead to increased pressure on the land. The result is the cultivation of steeper and steeper slopes involving clearance of biodiverse tropical forest that produces environmental problems in the lowlands, such as increased flash floods and the sedimentation of rivers and streams. The theory emerged in the 1970s and 1980s, when a few Western researchers began to take an interest in the environmental impacts of population growth in the rural parts of the developing world. For instance, in 1976, E. Eckholm wrote on Nepal:

> Population [increase] in the context of a traditional agrarian tech-
> nology is forcing farmers onto even steeper slopes . . . unfit for
> sustained farming even with the astonishingly elaborate terracing
> practised there. Meanwhile, villagers must roam further from their
> houses to gather fodder and firewood, thus surrounding villages with
> a widening circles of denuded hillsides.
>
> (1976: 77)

On the basis of this understanding of Himalayan environmental degrada-
tion, successive governments in the region have targeted upland farmers for over two decades. For instance, the Thai authorities announced a ban on all logging in 1989 and began a programme of reforestation involving plantations of teak, pine and eucalyptus. The ban and the reforestation policy have together altered the cultivation practices of upland farmers. Where the

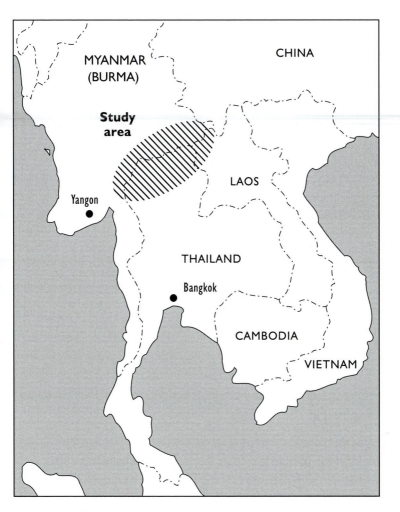

Map 3.1 Forsyth's study area

Karen practised rotational agriculture based around semi-permanent villages, and where immigrants typically cultivated plots for ten to twenty years before clearing new land, both groups now find their mobility restricted. In particular, there are government proscriptions on the cultivation of very steep slopes.

Forsyth is one of several environmental researchers who has sought to test the theory of Himalayan environmental degradation. His conclusion, as I shall explain momentarily, is that it is false and has been shown to be inaccurate for many years. Why, then, has it enjoyed such longevity and influence in countries like Thailand? His answer is that 'political and social factors' are the key (1996: 376). First, researchers like Eckholm, Forsyth argues, tended to arrive in the Himalayan region influenced by the neo-Malthusianism so rife in the West during the 1970s. They were, he argues, *already disposed* to see the transformation of the environment as unsettling a supposed 'balance of nature'. Second, in the particular case of Thailand, there is a long history of tension between lowlanders and uplanders in the north. In cultural, economic and political terms, lowland communities have been more central to Thai life than uplanders, who are still perceived as 'outsiders' because of their geographical distance from the rest of the country and their ethnic and linguistic difference from ordinary Thais. Finally, Forsyth argues that successive Thai governments have used environmental conservation in the uplands as a cover for their desire to gain military control over strategic high land near to the border with Laos and Myanmar.

If the theory of Himalayan environmental degradation is a myth then how can its mythical status be demonstrated? How can it be shown to be untrue and false?

ACTIVITY 3.3

Imagine you were in Forsyth's shoes. What sorts of things do you think you'd do if you wanted to dispute the theory of Himalayan environmental degradation? Would you contest it at the level of logic or facts or both?

The obvious answer to the Activity questions is that *empirical evidence* needs to be gathered that can test some or all of the following *testable hypotheses* in countries like Thailand:

(i) upland deforestation causes increased flash flooding and sedimentation in lowland areas;

(ii) upland inhabitants have cultivated land on steeper slopes as population pressure has increased;

(iii)   the cultivation of steeper slopes produces greater soil erosion and
        water run-off than on less steep slopes.

It's precisely this kind of evidence-based hypothesis-testing in which
Forsyth engages. In an essay entitled 'Science, myth and knowledge', he
presents research conducted in the village of Pha Dua, a settlement of
Chinese and Laotian immigrants located in Chiang Rai (Thailand's most
northern province). Settled in 1947, Pha Dua sits in an upland valley,
straddling steep granite, quartzite and sandstone slopes and a gentler valley
bottom. In 1995 it comprised 118 households and around 900 inhabitants.
The main cultivars are rice, maize and soya (grown on unterraced, non-
irrigated slopes), with an increasing preference for irrigated rice terraces
in the valley bottom. An influx of new residents, the establishment of
a government teak plantation, and the villagers' decision to preserve a
forested area for wood have together meant that agricultural land is scarce
in Pha Dua. Conducted in the early to mid-1990s, Forsyth's research was
triple-headed. First, he analysed aerial photographs of the village from
different decades in order to assess whether steeper slopes were being
cultivated over time. Second, he took measurements of the isotope Caesium-
137 that exists in soil profiles. By comparing these with measurements
taken from uncultivated soils, Forsyth could see whether erosion rates in
Pha Dua were significantly higher than natural rates. Finally, Forsyth under-
took a questionnaire survey of Pha Dua residents. He asked them about
their land-use practices and about their perceptions of land-use change in
Pha Dua over time.

When combined, the three sources of information could, according
to Forsyth, 'be used to falsify assumptions about Himalayan degradation'
(1996: 386–7). First, he discovered that Pha Dua farmers were aware that
soil erosion is higher on steep slopes which is why they preferred to culti-
vate flatter slopes more often as time went by. This was confirmed by the
analysis of aerial photographs. Second, Forsyth's isotope analysis suggested
that soil erosion in Pha Dua was no greater that in similar non-cultivated
areas. Though farmers in the village suffered from declining soil fertility
over time this was a function of nutrient removal by crops rather than
soil erosion. Indeed, an analysis of gulley-formation in Pha Dua suggested
that much of the sedimentation suffered in lowland areas of northern
Thailand might be natural: the result of gulleying on soils with granite
bedrock. As a result of his study Forsyth concluded that environmental

policy would do better to help farmers maintain soil fertility rather than focus on erosion-control measures. Overall, his research showed that it was wrong to see upland farmers as careless land-users. Though this may be true in some instances, the kind of careful case-study research in which Forsyth engaged is designed to question blanket statements about environmental degradation in the Himalayas.

In many ways, the kind of 'de-mythologising' approach to nature favoured by Forsyth takes us back to one of the original meanings of the word 'science'. When that word first came into currency – in seventeenth-century Europe – it had a very positive and progressive ring to it. It meant any kind of knowledge that was free from bias or prejudice and it was counterposed to the kind of religious, traditional and monarchical beliefs that permeated European society on the eve of the so-called 'Enlightenment' period (see Box 2.2 again). Today, of course, the word 'science' has negative as much as positive connotations – as I noted in Chapter 1 in my brief discussion of BSE. This is, no doubt, one reason why Forsyth prefers not to characterise his research as 'scientific'. Nonetheless, his insistence that evidence can adjudicate between myth and reality, fiction and fact, opinion and actuality places him in a lineage that, in geography, goes back to Harvey's critique of ideological representations of nature (see also Sullivan 2000).[3]

Hegemony and ideas of nature

If geographers like Forsyth seek to reveal the realities about nature concealed by false representations, others in the discipline are more concerned with how ideas of nature further the interests of ruling elites in various societies regardless of their factual in/accuracy. The focus here is less on the 'truths' hidden by these ideas and more on how the general acceptance of these ideas by those who, ironically, have something to lose by their acceptance is achieved. The Italian Marxist Antonio Gramsci called this process of acceptance 'hegemony' (see Box 1.4 again). As the British cultural geographer Peter Jackson (1989: 53) put it:

> hegemony refers to the power of persuasion as opposed to the power of coercion . . . [F]rom the point of view of a ruling class, . . . [this] is a much more efficient strategy than coercive control, involving the use of fewer resources and reducing the potential for

open conflict by securing the acquiescence of the oppressed to their subordination.

While hegemonic ideas serve the interests of dominant groups in society (e.g. men over women, ethnic majorities over ethnic minorities), they do not go uncontested. From time to time subordinate groups 'see through' these ideas, while dominant groups may seek to make new ideas hegemonic that unsettle older accepted ones. For instance, feminists have challenged the once taken-for-granted idea that 'a woman's place is in the home'. Meanwhile, many Western governments and businesses have promoted ideas of individualism over those of community since the 1970s in an attempt to weaken the power of trade unions whose traditional credo was 'all for one and one for all!'. Thus hegemony is a dynamic process in which dominant and subordinate groups battle it out to define which values, norms and beliefs will be the shared ones at any given moment in history. For Gramsci, power and resistance are not so much (or simply) physical acts as struggles over meaning. Importantly, Gramsci did not regard hegemonic ideas as false or misleading ones. Rather, they are partial or selective depictions of reality that appear to be otherwise because they are internalised as 'common sense' among the mass of the population.

Donald Moore's (1996) study of struggles over land use in the Kaerezi area of eastern Zimbabwe is a good example of how ideas about nature (in this case the environment) factor into the maintenance and contestation of hegemony. Moore, like Forsyth, is a political ecologist (based at the University of California, Berkeley). The element of his study I wish to focus on here is a 1990s dispute over the siting of a cattle dip in Kaerezi. Kaerezi is a rural area in a high rainfall belt that abuts Zimbabwe's Nyanga National Park, a major international tourist attraction. The cattle dip in question was established in 1988 by the national Ministry for Rural Development (MRD) for local livestock owners who were required by law to protect their herds from tick-borne disease. However, shortly thereafter another arm of national government – the Department of National Parks and Wildlife Management (DNPWM) – discovered that the dip was sited within 500 metres of the Kaerezi River. The area around the river had been designated a protected zone. To complicate matters further, a local trout-fishing club entered the fray, supporting the DNPWM's opposition to the siting of the dip near the river. What you had, then, was different parties clashing over the 'proper' use of a particular parcel of the natural environment.

What, it may be asked, has all this got to do with hegemony? Moore argues that what was at stake in this local dispute over a cattle dip were sets of ideas that were contested within Zimbabwe as a whole. Specifically, during the long period of colonial rule in Zimbabwe (or Rhodesia, as it was called) an attempt was made to supplant the ideas and beliefs of indigenous Africans with those of British society. In Kaerezi, land occupied by Chief Dzeka Tangwena was bought, without his permission, by white Rhodesians. Backed by the colonial state, white landowners in Kaerezi claimed that their paying for land superceded any rights Tangwena's followers might have by virtue of their occupancy and use of the land. This initiated a long period in which Kaerezi's black people were compelled to pay taxes and rents to white landowners and the colonial state. Added to this attempt to define property rights in land in terms of monetary purchase, part of Kaerezi was made into Rhodes Inyanga National Park in 1947 (the predecessor of Nyanga National Park). The park was a material expression of a growing British (indeed Western) belief that parcels of the natural environment were best preserved by keeping people out of them (on this see Neumann 1995, 1998 and Adams and Mulligan [2002]). Yet this belief clashed with the claims of Tangwena's descendents that the park was ancestral land that had been wrongly appropriated by the British.

The relevance of all this to the cattle-dip dispute, as Moore shows, is that the dispute was a local crystallisation of contests over the hegemony of colonial beliefs that surfaced strongly after 1980, when the British withdrew from Zimbabwe. The dip was not simply a hole in the ground filled with chemicals and the River Kaerezi was seen not simply as a water course. Rather, both were interpreted in terms of hegemonic ideas about property, conservation and rights to land. On the one side, parts of the post-colonial state (specifically the DNPWM) had internalised the conviction that environmental protection is best achieved by excluding people. On the other side, though, herders in Kaerezi had been granted property rights to land by the MRD, thus reversing the long history of colonial dispossession in the area. Meanwhile, the local trout-fishing club had a white membership, whose agitation against the dip struck black herders as a lingering example of colonial domination.

Faced with a divided state apparatus and local white opposition to the dip, Moore shows how livestock owners tactically used and opposed hegemonic ideas about property, conservation and land rights inherited from the colonial period. To quote him at length:

[A] local herdsman was quick to pick up on the white fishing club's claim, using a racial idiom to demand state action in defense of local residents' rights. Sitting on the tall grass amidst a circle of farmers, he pointed a bony finger at the [government] resettlement officer: 'Then *you* are the one who must go and fight these people. Why are invaders . . . coming into an area bought for people to settle on? . . .' When the resettlement officer countered that it was not the white club that controlled the land, but the government, the herdsman tactically concluded: 'So *you* want to kill the cattle? . . .' The state official then produced a letter voicing concern over pollutants from the cattle dip seeping into the river: 'This shows that the National Park was trying to take over the river, since the dip was already been approved by Veterinary Services', a department within yet another ministry.

(Moore 1996: 134)

In sum, Moore argues that hegemonic ideas instilled during the period of British occupation became the means through which local herders pursued their interests vis-à-vis the cattle dip. These ideas were both embraced and questioned, as white owners were bought out by MRD on herders' behalf and yet the DNPWM and angling club's arguments were resisted with reference to ancestral rights violated by the British. Tactically, herdsmen used their position of relative powerlessness to maximum advantage by manipulating representations of the Kaerezi. A critique of the imposition of British norms and values (like the separation of nature parks and people) was combined with a deliberate and pragmatic embrace of the British concept of landownership.

Discourse, nature and reality effects

The notion of hegemony directs our attention to how ideas of nature are battlegrounds where dominant and subordinate groups in any society confront one another. As Moore's research shows, geographical analysts of hegemony are interested in how reality is represented and with what consequences. Though they don't deny that hegemonic ideas refer to really existing things, these analysts are interested in the way these things are depicted rather than in their 'real nature'. This relative lack of interest in the biophysical realities of nature has been taken a step further by others in human geography. The third body of research into representations of nature that I consider in this section of the chapter questions the very distinction between ideas of nature and the phenomena those ideas refer to. In effect, this research collapses the 'gap' between representation and reality and subsumes the latter to the former. It does so by emphasising the power of discourse. As anthropologist Peter Wade (2002: 4) puts it, 'There can be no pre-discursive encounter with biology or nature'. The term discourse is one with many meanings in contemporary social science. At the most general level, it refers to a connected set of representations that 'regulate the production of meaning within . . . historically and socially specific situations' (Smith 2002: 343). Discourse-analysts in geography and other disciplines envisage societies as comprising multiple discourses that are sometimes contradictory, sometimes complementary. These discourses encompass cognitive, moral and aesthetic knowledge-claims and they specify what can (and, by implication, cannot) be known, said and done in any given situation. Discourses are directly linked to practice in so far as people act in accordance with the discourses they have internalised over time. For instance, consider the discourse of hygiene which is inculcated into all of us from a very early age by parents, schools, adverts for washing powder, the medical profession and so on. This discourse comprises a set of linked representations (such as: dirt = disease, cleanliness = civilised, odour = unattractive) that, in turn, inform the practices people perform on their bodies (e.g. regular showers, the wearing of laundered clothes etc.).

The idea that we both understand the world and act in it on the basis of myriad discourses differs from the notions of ideology, myth and hegemony in at least two ways. First, discourse analysts (as we'll see momentarily) question the idea that specific representations of the world serve the interests of definite groups in society. Instead, they see discourses as, if you like, impersonal 'grids' that condition the thought and action of any and all people who are exposed to these discourses for long enough. Whatever the specific origin of these discourses, they are seen to take on 'a life of their own' over time and they change only slowly as new or rival discourses emerge to qualify and challenge them. They do not always directly 'map on' to the intentions or agendas of identifiable social actors. Second, discourse analysts insist that it is impossible to know reality in a non- or extra-discursive way. Indeed, some of these analysts argue that the familiar distinctions between thought and matter, representation and reality, ideas and reality are *themselves a product of discourse*.

With this very general discussion of discourse in mind, we can identify four main versions of the idea current in human geography that nature is discursively constituted. Since I do not have the space to illustrate all four, I shall present case studies for just two of the versions discussed below (see Barrett [1992] for a good discussion of the notions of ideology, hegemony and discourse; most introductory books on 'cultural studies' also discuss these three concepts).

Cultures of nature

To begin, some human geographers have argued that discourses about nature are *culturally fabricated, culturally specific* and *culturally variable*. Here, discourses are more or less equated with the realm of culture. 'Culture' is an even more complex term than 'discourse' (and arguably as complex as the term 'nature'). Following the so-called 'cultural turn' in human geography (and several humanities and social-science disciplines) during the late 1980s, it has come to denote 'the medium through which people transform the mundane phenomena of the material world into a world of significant symbols to which they give meaning and attach value' (Cosgrove and Jackson 1987: 99). On this basis, some human geographers have identified the shared understandings of those things we designate as natural that are characteristic of particular societies. A good example of this kind of research is William Cronon's (1996) essay 'The trouble with wilderness; or, getting back to the wrong nature'. Cronon is a geographer and historian at the University of Wisconsin, Madison. Among his several research interests is a fascination with how the environment is interpreted in regions where different cultural groups come into contact. One of these regions is North America, where waves of immigrants (from Europe initially) displaced indigenous peoples from the seventeenth century onwards.

As its title suggests, Cronon's essay examines one of the most potent ideas in American culture, the idea of wilderness. I say idea because Cronon insists that wilderness is not what it appears to be, namely 'an area untouched by humans' (as the *Oxford English Dictionary* defines it). For him, wilderness is a culturally specific notion that has been applied to many natural environments in 'settler societies' like the USA, Canada and Australia. Of course, this claim that nature is a cultural construct rather than 'untamed nature' is a counter-intuitive one for many environmentalists. In the USA,

Plate 1 Wilderness: fact or fiction? This image of El Capitan and the Yosemite valley in the USA is readily identifiable as a 'wilderness' area: that is, an undisturbed area of natural beauty free of human habitation. However, it was not always so. Only through the efforts of various organisations – like the Sierra Club – and people – like photographer Ansell Adams – did the idea of wilderness take on its modern referents and signifieds (© Associated Press/Ben Margot)

for example, the belief that wilderness areas really exist underpins a good deal of green activism – like the opposition to George W. Bush's intention to exploit oil reserves in protected areas of Alaska during his presidency. So how does Cronon substantiate his argument that wilderness is a cultural construction whose cultural specificity is dissimulated?

ACTIVITY 3.4

We can begin to answer the above question by reflecting on what the term wilderness implies or connotes. When you hear the word wilderness what comes to mind? What meanings and values does the term conjure up? Jot down your answers as bullet points.

The obvious response to the Activity questions is that wilderness means parts of the non-human world (meaning 1 of the term nature) whose essence (meaning 2 of the term nature) is to be wild and unaffected by people. These cognitive meanings of the term aside, it also has naturalistic moral and aesthetic meanings (see Box 3.3). Morally, for many people there is something inherently good or positive about wilderness. It is readily contrasted with the stress and pollution associated with urban-industrial ways of life. Aesthetically, wilderness is also commonly seen as beautiful and uplifting because of its naturalness. Again, pejorative contrasts are often drawn with the crass or soulless appearance of towns and cities. Together, the moral and aesthetic connotations of wilderness help support a multimillion-dollar ecotourist industry in North America and beyond. This industry is based on the desire of (usually) urban dwellers to experience 'real nature' as they hike, camp, climb, ski or kayak in wilderness areas.

For Cronon, the layering of cognitive, moral and aesthetic meanings is too dense and complex to be a mere reflection of wilderness areas 'as they really are'. For instance, he claims it is no accident that wilderness gains its potency as an idea from a set of hierarchical contrasts that are semantically tied to it (nature versus society, rural versus urban, country versus city etc.). But the reason that wilderness is such a beguiling idea is that it appears *not to be an idea at all*. After all, who could doubt that a region like Alaska is relatively uninhabited and 'wild'? As Cronon (1996: 25–6) puts it,

---

Box 3.3 MORAL AND AESTHETIC NATURALISM

In Cronon's analysis (1996b) the idea of wilderness typifies what philosophers call moral and aesthetic naturalism. Such naturalism involves claiming that moral values and aesthetic beauty (or ugliness) can be 'read off' from those things societies categorise as natural things. In effect, this amounts to a 'nature knows best' argument where people are urged to learn their morality and their aesthetic values from the 'facts of nature'. Examples of moral and aesthetic naturalism abound in modern societies. For instance, homophobic people can often be heard to say that homosexuality is 'unnatural'. This claim implies that (i) heterosexuality is natural and *therefore* normal, and (ii) that any departure from heterosexuality is abnormal and *therefore* to be opposed, resisted and, if possible, eradicated. Here a judgement about homosexuality is *directly derived* from a supposed statement of fact about it, as if claims about the 'ought' (normative claims) can be mechanically determined by claims about the 'is' (cognitive claims) – see Saraga (2001). Moral and aesthetic naturalism denies that values are socially and culturally created. It is potentially authoritarian because it implies that our values are dictated to us by the natural world. By claiming that morality and aesthetics are scripted for us by the environment or our biology, such naturalism arguably conceals the specific values of the people who advocate it. This said, moral and aesthetic naturalism are not always rejected by left-wing thinkers and activists. If we take the case of gays and lesbians once more, we can see that it is potentially useful to insist that same-sex attraction is just as 'natural' as heterosexuality: that is, just as much a part of the 'way things are' and thus something to be accepted not stigmatised.

'Popular concern about [wilderness] . . . implicitly appeals to a kind of naïve realism . . ., more or less assuming that we can pretty easily recognise nature when we see it and thereby make uncomplicated choices between natural things, which are seen as good, and unnatural things, which are bad'. Against this, Cronon argues that the wilderness idea reflects the cultural values of increasingly industrialised and urbanised societies in which the natural environment seems to be fast disappearing. To quote him once more: 'The way we describe and understand the world is so entangled with our values and assumptions that the two can never be fully separated' (1996: 26). For Cronon, wilderness is almost a stereotype or ideal against which those who are anxious about the course of modern society evaluate that society. This is confirmed, Cronon argues, if we look at the history of the wilderness idea. For instance, in the pre-independence period, European settlers in the eastern USA saw wilderness as threatening, unruly and fickle – something to be conquered rather than embraced or protected (see Oelschlaeger 1991, Rothenberg 1995 and Nash 2001 for more on the idea of wilderness).

Not surprisingly, Cronon's views on wilderness have been seen as a provocation by several environmentalists in North America who tend towards the greener end of the ecocentric spectrum (see Box 2.3). These environmentalists have accused Cronon of being an anti-realist (denying the reality of the non-human world) and a moral and aesthetic relativist (supporting the view that all values are relative to the whims of a person, community or culture so that no values are better or worse than others – see Callicott and Nelson 1998 and Snyder 1996). I don't propose to assess the validity of these criticisms. However, I would point out that Cronon is not 'anti-wilderness'. His argument, rather, is that environmentalists need to be more honest about the source of their beliefs. For him, these beliefs do not emerge from wilderness but are imposed upon parts of the non-human world by certain cultures who have forgotten the particularity and constructedness of their values. In this sense, Cronon shows that wilderness is cultural 'all the way down'. For him there is no space outside cultural value systems in which areas like Alaska can be comprehended or evaluated. In the same spirit as Cronon, the cultural analyst Alexander Wilson (1992) has examined North American discourses about the environment, while Ramachandra Guha (1994) has offered a comparative perspective on the wilderness idea. More generally, the cultural critic Andrew Ross (1994) has examined how ideas of nature are always culturally saturated and specific, while Cosgrove and Daniels (see Box 3.2) pioneered geographical investigations into the cultural constructedness of landscape (one of nature's collateral concepts).

De-constructing discourses of nature

Cronon's understanding of discourse is, arguably, lacking in theoretical precision. Other human geographers, by contrast, have offered a more exacting understanding of how discourses operate. Derrideans, as the name suggests, take their understanding of discourse from the works of French philosopher Jacques Derrida (1930–2004). Derrida is usually classed as a post-structuralist, begging the question of what structuralism is (or was). In Derrida's case, the structuralism in question was that of the Swiss linguist Ferdinand de Saussure (1857–1913). Saussure's great influence as a theorist of language derived from the following claims. First, he argued that the relationship between words, meanings and things is entirely arbitrary. This is demonstrated by the fact that different languages use different words and sounds to denote the same things. Second, Saussure argued that meaning is produced within language rather than language mirroring an exterior social and natural world. Indeed, he's credited with identifying the signifier–signified–referent chain to which I referred in Chapter 1. Specifically, Saussure argued that all words and sounds take on a stable meaning in any society only because of their 'horizontal' and 'vertical' relationships with other words and sounds. For instance, consider the sentence 'The President's authority has diminished because his foreign policy has been economically costly'. According to Saussure we under-stand the meaning of this sentence only because (i) we understand how the meaning of each word is conditioned by its relations with the others in the sentence and (ii) we understand the absent synonyms and antonyms for each word (for instance, we could substitute the words 'elected leader' for President without changing the meaning of the sentence). For Saussure, then, 'language does not map on to pre-existing differences out there in the world, but creates those differences' (Edgar and Sedgwick 2002: 209). Finally, this led Saussure to conclude that language is a system or structure with definite rules that control what can and cannot be said at any given moment in time. Just as rules of a game circumscribe the moves of players, so those of language (*langue*) frame the particular utterances (*parole*) of interlocutors.

Derrida, whose 'de-constructive' writings began in the 1960s, took Saussure's structural linguistics a step further. Derrida argued that if all reference is arbitrary and if meaning is generated within linguistic systems, then it is impossible to establish final or correct representations of anything.

For Derrida all meaning is created through a simultaneous process of difference and deferral that is internal to language. For instance, the meaning of the word 'nature' in any given context depends upon its opposition to terms like 'culture' or 'society'. These latter terms are nature's 'constitutive outsides' or 'absent presences': the word nature *needs* these antonyms in order to be understood. For Derrida it follows from this that the meaning of a word or sound is thus always deferred (or postponed). Since meaning is never wholly present in the word or sound used then the apparent stability of meaning is only ever that: *apparent* not real. The intellectual project of Derrida and his followers has thus been to de-construct language. Derrideans show how apparent certainties of meaning are always subverted if one subjects them to a 'symptomatic analysis'. This is not to say that there is not *relative stability* in the relationship between signs, meanings and referents. What Derrideans are saying is that this relationship is not given in nature but, rather, contingent and open to challenge. It should be noted that in Derridean circles, the terms 'language', 'text', 'representation' and 'discourse' are often used interchangeably.

This capsule description of Derrida's thinking is sufficient to help us comprehend a germinal piece of research by Bruce Braun, a University of Minnesota geographer. Braun has combined Derrida's ideas with those of post-colonial critics to analyse how a high-profile 'wilderness area' has been understood by those groups most concerned about its fate. I mentioned post-colonial thinking in passing in Chapter 2. Here it's sufficient to say two things about this mode of thinking. First, following Said's (1978) classic book *Orientalism*, it argues that colonialism is still with us − even though the era of formal colonial occupation by Western powers may be at an end. Specifically, post-colonial critics argue that colonial beliefs about non-Western Others still infuse Western cultures. This means, second, that the term 'post-colonial' is simultaneously literal and ironic. Though post-colonial critics acknowledge that Western countries no longer rule in Africa, Asia and elsewhere, they question whether we are quite as 'post-' (or beyond) the colonial period as we think we are. The wilderness area the struggle over which Braun has examined is Clayoquot Sound in British Columbia, Canada. His analysis is more precise than Cronon's, though it has the similar intention of exposing how the non-human world cannot ever be comprehended 'in the raw'. How, then, does Braun scrutinise discourses of nature in Clayoquot Sound using Derridean and post-colonial thinking and to what end?

Clayoquot Sound is an ocean inlet containing increasingly rare stands of 'old-growth' temperate rainforest (see Map 3.2). The area came to international attention in the early 1990s when environmentalists in British Columbia strongly opposed the granting of a logging licence to the forest-products multinational Macmillan Bloedel. Though the British Columbian economy depends upon timber exports to a considerable extent, it is also the place where Greenpeace came into existence (in 1971), while Vancouver – the province's largest, most cosmopolitan city – is home to many people with strong environmental sensibilities. In his essay 'Buried epistemologies: the politics of nature in (post-)colonial British Columbia', Braun (Willems-Braun 1997) de-constructs the discourses used by environmentalists and the pro-logging lobby respectively in their depictions of Clayoquot during the early 1990s. Specifically, he focuses on key publications produced by both sides of the dispute in their attempt to win over the government and public of British Columbia. At first sight, these publications depict the same forest (Clayoquot's old-growth trees) in very different ways. The Macmillan Bloedel document, entitled *Beyond the Cut*, mixed glossy photographs, text and graphics in an easy-to-read format. Braun shows that the company depicted Clayoquot's trees as a valuable *resource* that belongs to the Canadian nation. Having chosen this dispassionate language, Macmillan Bloedel constantly emphasised its credentials as a responsible resource-manager – a custodian of the forest on behalf the Canadian people. Overall, Braun shows, *Beyond the Cut* positioned Macmillan Bloedel as an experienced and ethical company keen to create jobs for people in the forest industry while being careful to cut down trees in an environmentally responsible way (see Plate 2).

Not surprisingly, environmentalists' depictions of Clayoquot departed somewhat from the Macmillan Bloedel view. Drawing upon the wilderness idea that Cronon dissects, these environmentalists represented Clayoquot in altogether more emotive and ecocentric terms. Braun focuses on *Clayoquot: On the Wild Side*, a very popular coffee-table book published by the Western Canada Wilderness Committee (which was actively involved in opposing Macmillan Bloedel's logging practices in the early 1990s). Unlike *Beyond the Cut*, the book's story is told much more in visual terms (160-plus images by nature photographer Adrian Dorst). The photographs depict Clayoquot as 'a sublime, complex, enchanting landscape filled with powerful forces and intricate, even delicate, relations' (Willems-Braun 1997: 19). Most of them focus on the natural world (particularly

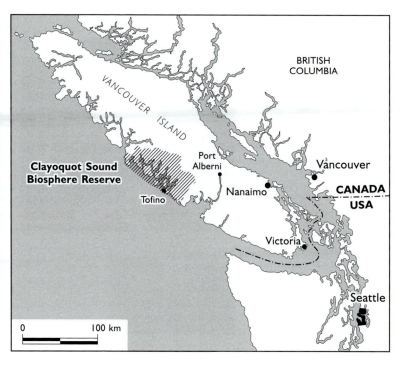

Map 3.2 Clayoquot Sound

Clayoquot's colossal trees) and very few contain obvious signs of a human presence. Where they do, as Braun points out, people are either dwarfed by the enormity of the natural landscape (see Plate 3) or else made to appear 'in harmony with it'. In the latter case, several images of an indigenous group local to Clayoquot (the Nuu-chah-nulth) are included in the book. As Braun notes, these images depict natives as an unintrusive presence 'at one' with a natural environment central to their traditional way of life. In contrast to the scientistic discourse of timber resources and their rational management, *Clayoquot: On the Wild Side* represents the area as a majestic landscape whose value lies not only in the rarity of its trees but also its undespoiled character.

So far so good. From my summary thus far, it may seem as if Braun's analysis is a species of ideology criticism, de-mystification or hegemony-critique as per my discussion in previous sections of this chapter. However, this is not the case. Braun's point is that despite their apparent differences,

■ FORESTRY RESEARCH

Forestry research includes silviculture and land use. The focus within silviculture is three-fold:

■ REGENERATION

The development of techniques which best prepare growing sites and establish the next crop to ensure its survival.

■ GROWTH AND YIELD

The development of a database to realistically predict the rate of growth and the expected yield from the wide variety of growing sites. Soil conditions, weather, elevation, animal habitat are considered.

■ FERTILIZATION

Research-based decisions on the "how" and "when" of fertilization to optimize the potential for growth and yield improvement of the forest.

Landuse research concentrates on the impact of forest management practices on soil, fisheries and wildlife resources. This work contributes to the integrated management of the company's land base giving consideration to the variety of needs and demands placed on it.

Plate 2 Clayoquot as a natural resource. This page of a Macmillan Bloedel pamphlet on 'scientific forestry' depicts Clayoquot Sound as an abstract stock of timber to be rationally managed on behalf of Canadians by 'responsible' forest companies like MacBlo (Reproduced from Macmillan Bloedel Research)

the pro-logging lobby and environmentalists represent Clayoquot in similar ways. In other words, their *different* intentions and aspirations for the forest aside, both groups unwittingly deploy the *same* discourse about the non-human world (albeit in rather different idioms, one dispassionate, the other romantic and moralistic). According to Braun it is a colonial discourse that supposes that the 'true' character of Clayoquot is its naturalness (i.e. the absence of a human presence). It is colonial because it erases (or minimises) the presence of indigenous peoples. In *Beyond the Cut*, Braun argues, Clayoquot is depicted purely as a stock of valuable timber, while

Plate 3 Clayoquot as wild nature. This photograph by Adrian Dorst depicts Clayoquot as an ancient natural landscape of majestic trees and intricate ecological relationships among plants, mosses and fauna (© Adrian Dorst)

in *On theWild Side* it is displayed as a pristine wilderness. Both representations appear to represent a mute nature (trees) that cannot represent itself. Following Derrida, Braun argues that this appearance is achieved by virtue of the semantic oppositions of nature–culture and traditional–modern. The framing of Clayoquot as a natural space lightly peopled by natives who respect the environment is, he argues, *an effect* of a culturally specific discourse that goes back to the arrival of the British in the 1850s. This discourse claims to represent nature as it is, but in fact gains its meaning only from a set of dualisms internal to the discourse itself. Each side of these dualisms refers 'sideways' to its antinomy rather than arising 'vertically' from its real-world referent. Over a century after the British colonised the province and decades after Canada ceased to be a British dominion, Braun argues that colonialism lives on – in this case in struggles over the environment. The British placed indigenous peoples in 'reserves', constraining their occupancy of forests, valleys and mountains in British

Columbia. Even in our supposed post-colonial period, Braun shows that this confining of natives to circumscribed spaces is repeated in the representations of Macmillan Bloedel and environmentalists. These representations acknowledge neither the history of displacement of native peoples nor their contemporary claims to own and use Clayoquot and other parts of Canada. In sum, Braun (2002: 17) shows that 'what counts as nature cannot pre-exist its [discursive] construction' and that labelling some things as natural is politics by other means (see also Braun 2000 and Braun and Wainwright 2001; for more on Clayoquot, see Magnusson and Shaw 2003).

Discourse, discipline, nature and Foucault

The third way in which discourses of nature have been understood by critical geographers takes us away from the study of the non-human world and more towards 'human nature': that is, the mind and the body. As I mentioned towards the end of Chapter 2, those on the 'cultural left' of human geography have made concerted efforts to de-naturalise our understanding of subjectivity/identity and corporeality. In many, but by no means all, cases, these researchers have drawn upon the germinal ideas of the philosopher and historian Michel Foucault (1926–84). Foucault's many writings contested the notion that minds and bodies are given in nature. Though he recognised that humans are born with neurological and corporeal capacities that distinguish them from other species, he saw these capacities as surfaces upon which discourses worked slowly but steadily over time. In other words, Foucault argued that people's mental and physical characteristics were largely societal products not prescripted by their biological make-up.

Foucault saw societies as comprised of multiple, overlapping and often conflicting discourses. In contrast to the Marxian notions of ideology and hegemony, Foucault argued that discourses do not serve the interests of one or other dominant group in society. Rather, he saw them as more diffuse and anonymous. In his historical studies of discourses about madness, criminality and sexuality, Foucault showed how discourses constitute the phenomena they purport merely to represent. For instance, his History of Sexuality (1979) traces the changing ways in which sexual identity and sexual behaviour has been understood in Western societies. As its title suggests, this book shows that sexuality is changeable and inconstant over time, not a timeless biological imperative. In terms of sexual identities, the book shows how individuals' sexual self-understanding is, in large part, assigned to them by the discourses of sexuality current at any one moment in time. These discourses offer people a limited range of 'subject positions' that they can inhabit – like straight, gay or bisexual. People are 'recruited' into these positions rather than choosing them for themselves. What is more, Foucault, argued, these subject positions are always already judgemental and prescriptive. Thus, those individuals who consider themselves to be heterosexual are regarded (and regard themselves) as 'normal', while people who are labelled 'homosexual' have, historically, carried the burden of being classified as 'abnormal', 'deviant' and even 'perverse' (see Box 3.3). In terms of bodily practices, Foucault showed that discourses of sexuality were not only internalised mentality. More than this, they become *practised* discourses because individuals come to regulate their sexual habits in accordance with these discourses. These habits become 'routinised' through repetition throughout a person's life-course. As Foucault (1979: 105) put it, 'Sexuality must not be thought of as a kind of natural given which power tries to hold in check, or as an obscure domain which knowledge tries gradually to uncover' (see Segal 1997 for more on the different ways that sexuality has been understood in Western societies).

For Foucault, then, discourses of sexuality and much else besides 'take hold' of people's identity and their bodily comportment. He saw subjectivity and corporeality as malleable across time and space – as enmeshed in well-established discourses that are propagated through major institutions such as schools, prisons and hospitals. Though in his later work he explored how resistance to discourse might work, in much of his writing Foucault saw discourse and power as coterminous. After all, if people cannot find a space outside of discourse then they can only think and behave in ways that are *already established for them* (not *by them*). Power is thus not, for Foucault, something that emanates from, or is held, by key societal institutions (like the state). Rather, it emanates through them in a 'capillary' fashion as multiple discourses do their daily work on people's minds and bodies. Power is thus 'productive' in much of Foucault's theorising: it *creates* the kinds of conforming identities and modes of behaviour that are commensurate with dominant discourses.

This summary of Foucault's ideas is inevitably inadequate. But it is detailed enough to make the link with several geographers' research into people's subjectivity and bodily action. Though Foucault never wrote about nature in the formal sense, the concept of nature and many of its collateral concepts (like sex and sexuality) are integral to the discourses he analysed. In particular, Foucault showed that in many societal discourses the natural and the normal are equated, as are the 'unnatural' and the 'abnormal'. For instance, as noted above, if a heterosexual identity and associated sexual practices are seen as 'normal' in a society, then a person who is homosexual may well feel the need to conceal both their sexual preferences and their sexual behaviour. The geographical link here is that all dominant discourses are reproduced in and through myriad physical sites that are arranged in such a way as to reinforce (or challenge) these discourses. For instance, the British 'queer geographer' Gill Valentine (1996) has examined the heterosexual coding of public space and the effects of this on the time–space behaviour of homosexuals. Sexuality aside, other geographers, such as Chris Philo of Glasgow University, (Philo 2001) have used Foucault's ideas to explore the geographical constitution of discourses of sanity and insanity – where, once again, notions of nature are in play (since insanity is often thought to be a congenital illness residing in an individual's head – a case of 'mental functioning gone wrong').

Foucault is not the only major thinker to have inspired human geographers in their de-naturalisation of mind and body. Others include Judith Butler, Julia Kristeva, Jacques Lacan and Pierre Bourdieu (see Lechte 1994 and Edgar and Sedgwick 2002 for pithy introductions to these thinkers' work). For instance, the British cultural geographer Steve Pile (1996) has used insights drawn from psychoanalytic theory to explore how different identities are fashioned in urban environments. Using Foucault and other theorists of identity and corporeality, critical human geographers have de-naturalised gender, 'race' and sexuality in particular. I've focused on Foucault because he brings identity and the body within one analytical framework and also because discourse is so central to his thinking. It should also be noted that, unlike Derrida, Foucault's work evinces a highly materialistic understanding of discourse. Because discourses 'reach in' to the mind and body they have a palpable physicality. In this sense, the research of Foucault and those inspired by him shades into the material constructionism dealt with in the section 'Remaking nature' below.[4]

Hyperreality and virtual natures

The fourth, and final, approach to discourses of nature that I want to focus on is partly inspired by a contemporary of Derrida and Foucault. Jean

Baudrillard (1929–) – a philosopher, cultural critic and media theorist – has famously argued that we increasingly live in a world of simulation. As the English sociologist Mark Smith (2002: 287) puts it:

> Baudrillard argues that it is no longer possible to distinguish between representation and reality in a conventional way, one which assumes that representation refers to something which really exists. For Baudrillard, all kinds of representations are just 'simulations' of the meanings which have been produced before. This condition, which he describes as *hyper-reality*, involves the blurring of the distinction between . . . 'fact' and 'fiction'. In hyper-reality meaning is not produced but reproduced through simulations, and simulations of simulations and so on.

In Baudrillard's work, then, the term discourse refers to that complex array of images, words, texts and sounds through which people communicate with one another about the world in which they live. It is, if you like, a generic term not just for language but for any organised sets of signs and symbols which shape our understanding of reality. In the modern world, Baudrillard argues, people's experience of reality is increasingly indirect and mediated. Television, video games, movies, theme parks and the like are, for Baudrillard, the 'reality' that more and more people inhabit on a daily basis. In his later writings, Baudrillard has been preoccupied with how gatekeepers of knowledge (like CNN and the BBC) mediate the experience of the world for society as a whole. For instance, his ironically titled *The Gulf War Did Not Take Place* (1995) argued that the Gulf War experienced by citizens of the allied-forces countries was a media construct. This media war, Baudrillard argued, drew upon a well-established repertoire of representations of Arabs which meant that 'representations of reality . . . precede[d] that reality [and thus] cease[d] to be representations and bec[ame] . . . simulations instead' (McGuigan 1999: 61).

The British geographers Rob Bartram and Sarah Shobrook (2000) have drawn upon Baudrillard's ideas and applied them to the experience of nature in Western societies. Taking the case of the Eden Project, they insist that what we often think of as 'first nature' – that is, nature untouched by human hand – is really a simulated nature that is anything but real. Like Cronon's intervention, the context for their argument is the heightened anxiety about the 'end of nature' in Western societies (refer back to the penultimate nature story in Chapter 1). The Eden Project, located in the county of Cornwall in England, is, Bartram and Shobrook argue, symptomatic of this anxiety. The project was constructed at a cost of £74 million as part of Britain's millennium celebrations. At first sight, it is an attempt – one of many worldwide – to forestall the disappearance of the non-human world bequeathed by evolution. The project's biblical name intentionally invokes the image of a pre-modern time in which people lived in harmony with the environment – an image which, for the project's creators (professional conservationists), stands as a criticism of present-day environmental abuse. Focused on plant life in particular, the project's geodesic-dome greenhouses recreate biomes from around the world (humid-tropical, warm temperate etc.). With meticulous attention to detail, each biome contains the species found in the equivalent real-world ecosystem. Using the latest technology the climate inside the domes is carefully controlled. Tourists are able to walk through each greenhouse, experiencing in microcosm the biomes that, in the real world, are threatened by human activities like logging and road-building. In effect, the project aims to educate the public about the intricacy and value of the natural environment by giving them direct access to that environment. The project's many visitors would otherwise have to travel around the world in order to gain first-hand acquaintance with the biomes recreated in the various greenhouses.

Bartram and Shobrook refuse the obvious interpretation of the nature that the Eden Project presents to the public. Instead of arguing that the project's biomes are capsule versions of real biomes – complete and authentic in almost every detail – they argue that it is a simulated or virtual nature. Following Baudrillard, this means that the plant life assembled in the geodesic domes do not 'stand for' or represent 'real plant life' in various parts of the world. Instead, argue Bartram and Shobrook (2000: 371), it stands for 'past events, images and ideas' about the environment that reflect Westerners' long-standing anxiety about the disappearance of non-human species. 'Paradoxically', Bartram and Shobrook write, 'the closer we get to the real world [in the Eden Project], the more detached and remote we become from it' (2000: 372). Though tourists' experience of nature in the domes *seems* to be direct, visceral and first hand, it is arguably anything but that if one follows a Baudrillardian logic. As Bartram and Shobrook insist, the project's biomes are part of a scripted morality play into which tourists are inserted. Symptomatically, the biomes are free of the pests and diseases one finds in their natural equivalents. Bartram and Shobrook

Plate 4 The Eden Project: nature protection or projection? These geodesic domes house 'natural biomes' that are intended to reproduce their real brethren. But are the biomes faithful copies or virtual realities? (© Science Photo Library)

see this not simply as a practical necessity but as a cultural metaphor wherein order, purity and innocence is valued more highly than disorder, disease and degradation. Visitors to the Eden Project thus arrive having *already* internalised a set of cultural beliefs about 'first nature' that the Project merely *confirms and reproduces* with its seemingly 'realistic' recreation of natural biomes.

It's important to note that Bartram and Shobrook are not arguing that the nature recreated in the Eden Project is a 'fake'. Rather, they are suggesting that the very distinction between the fake and the real, the representation and the reality has disappeared. For them, what we call 'reality' is an effect of discourse; it is a distinction *internal* to discourse even though it appears to be a distinction *between* discourse and a world external to it. Sceptical readers might object that Bartram and Shobrook are wrong, because it is possible to check the authenticity of the project biomes by comparing them with the real biomes they mimic. However, the counter-argument is that one would view the supposed 'real biomes' with the self-same cultural beliefs that, Bartram and Shobrok argue, make the Eden Project a simulation.

This argument has been made powerfully in another context by the historian and cultural theorist Timothy Mitchell. Influenced by Baudrillard (as well as Derrida and the German philosopher Martin Heidegger), Mitchell's (1988) *Colonising Egypt* examines how the French and British viewed Egypt during the late nineteenth century. He recounts how, at the World Exhibition of 1889 in Paris, a 'faithful' recreation of an area of old Cairo was built for the edification of the middle-class public. Compared with British and French streetscapes, the Cairo one was dirty, rather chaotic and traditional in appearance. It comprised winding, irregular alleyways and overhanging façades. For those exhibition visitors who had been to Egypt before, the artificial Cairo street appeared authentic, while for those visitors who hadn't it seemed equally realistic because it was built to scale and was so obviously different from the typical British or French streetscape. What's startling about Mitchell's analysis is his claim that even those who had (or would in future) visit Cairo *never stepped outside a set of cultural representations of Egypt*. These representations traded on a set of binary oppositions between West and East, light and dark, order and chaos, cleanliness and filth, civilisation and barbarism, that structured how Western Europeans saw not only Egypt but the developing world at this time. Thus, the 'reality' that the artificial Cairo street at the exhibition stood for was not, in fact, a street in old Cairo but an image or prejudice that the putative reality seemed to confirm. As Mitchell (1988: 10) put it, 'despite determined efforts within the exhibition to construct perfect representations of the real world outside, the real world beyond the gates turned out to be rather like an extension of the exhibition'. To bring things back to Bartram and Shobrook's work, the nature depicted in the Eden Project can, for them, be seen not as a first nature nor even a 'second nature' (i.e. one physically modified by human intervention) but, rather, a 'third nature' (Wark 1994). This third nature is purely discursive and collapses any divide between simulation and reality (and it relates to the 'non-representational theory' I discuss in Chapter 5: see Smith 2003).

REMAKING NATURE

The argument that nature is 'constructed' at the level of representation may seem to ignore the biophysical world to which representations of nature refer. A second main branch of constructionist research in human and environmental geography focuses precisely on this world. This research is preoccupied with *material constructionism*: that is, the process whereby societies physically reconstitute nature so that it is no longer natural. This is the second kind of social constructivism Demeritt (2002) identifies in his review of human geographers' recent research on the topic. In its strongest form, this research argues that nature is a physical construction through and through so that it makes little sense to call it 'natural' any more. In spirit, if not always in theoretical substance, it builds on the earlier work of people like Hewitt and Watts which, as we discovered, argued that the physical capacities of nature must always be defined *relative to* specific forms of societal organisation. In this section of the chapter I identify two main ways in which the biophysical capacities of nature have been taken seriously by contemporary human and environmental geographers (see Bakker and Bridge 2003 for a more refined discussion). As we move to the second we see that these capacities are regarded as *ontological products of social relations, processes and actions*. In order to focus the discussion, I concentrate on research by Marxist geographers in the subfields of agro-food studies (or what used to be called agricultural geography) and natural-resource analysis. This is not the only research into nature's physical reconstitution in human and environmental geography (see Box 3.4). But it does have the virtue of being one of the major strands of inquiry into the topic.[5]

Box 3.4 CRITICAL GEOGRAPHY AND THE MATERIAL CONSTRUCTION OF NATURE

Aside from Marxist geographers who study the transformation of the physical environment there is at least one other cluster of left-wing human geographers preoccupied with material constructionism. So-called 'environmental injustice' researchers examine the distribution of 'environmental bads' (like toxic waste dumps) vis-à-vis poor and marginalised groups of people. Here environmental hazards are seen as ones actively (though often unintentionally) created by industrial societies. Aside from environmental injustice researchers and Marxists, there are other human geographers who are interested in the material impacts of the non-human world upon people. These geographers would not necessarily describe the non-human world as a physical 'construction' but neither do they think it is a domain wholly separate from human intervention. An example is the research of British cultural geographers Jacquie Burgess and Carolyn Harrison (1988) into everyday attachments to urban parks and green spaces. Burgess and Harrison acknowledge that these parks and spaces have been designed by planners, but they also show how the material properties of these sites – in terms of colour, smell and physical layout, for example – matter immensely to how local residents use and value them. Likewise, many Third World political ecologists show how the physical environments produced through human action have an active role to play in both enabling and inhibiting the needs of different individuals and groups at the local and extra-local levels.

Intellectually, this Marxist research has emerged in the wake of an important essay written by the University of California geographer Margaret Fitzsimmons. In 'The matter of nature' (1989), Fitzsimmons reprimanded left-wing geographers for having ignored the ways in which societies are transforming the physical environment. Fitzsimmons insisted that the biophysical properties of the non-human world 'mattered' in the double sense that they were materially important for societies and should, therefore, be topically important for human geographers. The research discussed in the two subsections below focuses on economic sectors where the physical environment is 'confronted directly': i.e. the extractive sector (mining and fisheries) and the cultivation sector (agriculture, aquaculture and forestry).

The materiality of the non-human

Regardless of how it is represented, the non-human world has definite material characteristics that present both obstacles and opportunities for societies – or so many present-day Marxist geographers would argue. These physical characteristics constrain and enable how different societies use the environment. At issue here is not environmental determinism – where societies become a passive, dependent variable – but, rather, a *society–environment dialectic*. A dialectic is a dynamic, two-way relationship of mutual influence and adjustment. As I noted in my earlier discussion of Michael Watts's research, Marxist geographers are especially interested in how capitalist societies appropriate natural resources. In other words, they are interested in the *historically specific* society–environment dialectic found in capitalist societies which, in turn, *varies geographically* depending on the specific elements of economy and environment in question. For many Marxist geographers, the non-human world still has a degree of relative independence and agency – even in an era where it seems that we can control that world down to the genetic and molecular level (see Hudson 2001: ch. 9). William Boyd, Scott Prudham and Rachel Schurman (2001: 557) capture well the challenge of studying this world from a geographer's perspective: 'On the one side lies the danger of overlooking the significance of the biophysical world in nature-based industries and lapsing into pure social constructionism. On the other lies the spectre of environmental determinism'.

This negative comment about 'pure social constructionism' indicates a determination on the part of some Marxist geographers to avoid seeing the non-human world as mere putty in the hands of modern societies. Indeed, within the Marxist academic community more generally (which cross-cuts several disciplines), a process of 'greening' has occurred over the past decade or so. This has involved taking seriously the way that the non-human world physically conditions what societies can (and cannot) practically do. It combines an attention to the internal structure of capitalist societies with an equal attention to the material characteristics of the resources and environments upon which those societies depend for food, shelter, warmth and much else besides. As Boyd et al. (2001: 557) put it: 'While the social constructivists are right to argue that [the environment] . . . should not (indeed cannot) be de-historicized (i.e. placed in a category outside of human history . . . and social relations), we agree with those who argue that there *is* a material "other" to [social] . . . processes'.

An example of this kind of 'both/and' research that sees the environment as neither wholly autonomous nor wholly a product of social processes, is my own investigation into commercial sealing (Castree 1997: 1–12). Though an historical study (it focuses on the overexploitation of north Pacific fur seals in the late nineteenth century), it should not be thought that the intellectual framework deployed has no contemporary relevance. The studies of Gavin Bridge (2000) and Roberts and Emel (1992) indicate as much, since they examine present-day copper-mining and ground-water extraction using a Marxist framework not a million miles from the one used in my 1997 essay. I shall discuss this framework momentarily, adding to my earlier discussion of Marxism in relation to the writings of Harvey and Watts. But let me start with a summary account of the so-called 'war against the seals' in the Bering Sea between 1870 and 1911.

In less than forty years, four sealing fleets and two land-based sealing companies located in eastern Russia, Japan, western Canada and California virtually exterminated the north Pacific fur-seal population. This population was prized because the pelts of fur seals could be made into warm and especially luxurious garments – like winter coats and capes. The market for these garments was very lucrative and involved fashion-conscious middle-class consumers in cities like London, New York, Paris and Moscow. The main players in the fur-seal trade were two companies based in the Pribilof Islands (islands which belonged to the USA after 1867 as part of the Alaska purchase), and four sealing fleets (two on each side of the north Pacific ocean). The reason so many players were involved is because fur seals are naturally mobile (a so-called 'fugitive resource'). Each year they migrate through the territorial waters of eastern Russia, Japan, British Columbia and the western USA, pausing for two months on the Pribilof Islands to give birth to offspring and to mate (see Map 3.3). When the USA purchased Alaska, a group of San Franciscan financiers realised that a good deal of money could be made killing seals on the Pribilofs and were granted a twenty-year licence to do so by the US government of the time.

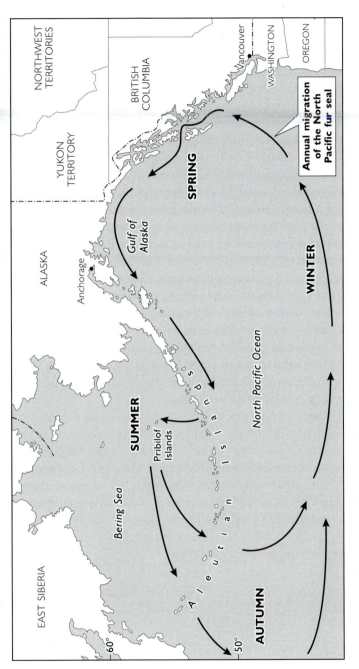

Map 3.3 The migratory path of the North Pacific Fur Seal

Subsequently (from 1890) a second company was awarded this exclusive land-sealing licence. Between 1870 and 1910, the two American land-based sealing companies averaged profits of over 100 per cent – a fact which indicates just how lucrative fur-sealing was at this time. During the same period, settlers in the new province of British Columbia realised that they too could enjoy the economic rewards of sealing by shooting or spearing fur seals as they swam along the coast in early spring each year. Meanwhile, the Russians and Japanese – who had been sea-sealing for generations – continued to take a share of the fur-seal population once the seals left the Pribilof Islands in late summer. By the early twentieth century, sealers in the four countries had reduced 3.5 million seals (in 1870) to just a few tens of thousands. The near-decimation of the fur-seal population was arguably the first international 'environmental problem' of the twentieth century. It almost brought the countries involved to war, as each argued that it was entitled to harvest seals without restriction. The problem was only solved after several years of diplomacy, culminating in the North Pacific Fur Seal Convention of 1911. This precedent-setting agreement sought to protect the seals while making money off their slaughter in a controlled way. Specifically, all sea-sealing was banned because sealing on the Pribilofs (controlled by the USA) was the only way to count accurately how many seals were being killed each year. In return for losing their sealing fleets, Canada, Japan, Russia and California were given a fair percentage of the money made by Pribilof sealing each year.

From a Marxist perspective, the virtual destruction of the seal herd by 1911 exemplifies the liability of capitalist societies to destroy their own natural-resource base. In other words, it illustrates the fact that capitalism is an 'ecologically irrational' economic system which generates environmental problems as part of its 'normal' functioning. Let me explain. Earlier in this chapter I discussed the Marxist concept of the mode of production. A mode of production, you will recall, is the particular way a given society (or societies) produces goods and services. In capitalist societies, goods and services are produced in the following way:

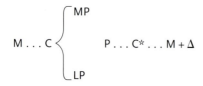

$$M \ldots C \begin{cases} \text{MP} \\ \\ \text{LP} \end{cases} \quad P \ldots C^* \ldots M + \Delta$$

Here, a firm advances a sum of money (M) to purchase commodities C – namely, MP (means of production: raw materials, machinery etc.) and LP (labour power: people's capacity to work). When combined in a process of physical production (P), MP and LP yield new commodities (C*) that are sold for the original amount put forward plus an increment (Δ: profit). This profit is, in turn, reinvested in the next round of production and so on in a spiral of growth. Among other things, a capitalist mode of production is distinguished from others by the following three characteristics. First, there are two main productive classes (i.e. owners of firms and those who work for them). Second, the principal aim of production is profit (what Marx called 'growth for growth's sake') rather than, say, a decent standard of living for all. Capitalist firms are primarily interested in the exchange value of the commodities they produce rather than their practical use value. Finally, capitalist firms relate to one another competitively: they have to fight for market share.

What has this got to do with the north Pacific fur-seal case? First, from the 1870s, seals were valued in monetary terms rather than, say, moral terms. Commensurate with the production goals of capitalist societies, they became mere vehicles for the realisation of profit on the part of sealers. Rather than being seen as inherently good or valuable they became prized for the exchange value of their pelts. Second, because sea-sealers and the land-based companies were locked into relations of competition it was 'rational' for them to kill seals willy-nilly – even though they knew that one day the seal herd might be decimated. This was exacerbated by the fact that the sealers were divided by nation, reducing the chance of a more controlled, cooperative approach to harvesting seals. For these two reasons, it is no surprise that the seal herd was badly overexploited by the early twentieth century. However, this overexploitation was not only due to the anti-environmental 'logic' of capitalism. It was also due to the 'nature' of the fur seals.

ACTIVITY 3.5

On the basis of the Marxist analysis presented in the previous two paragraphs, in what ways did the fur seals 'matter'? Make a list of the ways the seals' natural characteristics shaped the fur-seal trade in the north Pacific.

Following Fitzsimmons (1989), the word 'matter' in the Activity question has a dual meaning. The question gets you thinking about how the physical nature (matter 1) of the fur seals made them important (matter 2) to the four countries involved in the fur trade throughout the 1870–1911 period. A useful way to answer the question is to distinguish the *possibilities* and the *obstacles* that fur seals presented to both sea- and land-based sealers by virtue of their physical constitution. In simple terms, the possibilities were as follows: (i) the dense pelts of seals created a market for garments made from these pelts; (ii) the large number of seals (3.5 million in 1870) made commercial sealing a viable prospect; (iii) the fact the seals congregated on land for two to three months per year made sealing attractive for the two US companies; (iv) the fact that the seals migrated through the north Pacific for nine to ten months per year made sea-sealing attractive for Canada, Russia, Japan and California sealers. In terms of obstacles, the following were arguably important: (i) for obvious reasons, it was difficult to count seals in the ocean, meaning that sea-sealers were never sure of the ratio of killed to living seals; (ii) likewise, it was difficult to determine the sex of seals at sea, meaning that many pregnant seals were accidentally killed, thus undermining the reproductive capacity of the seal herd over time.

In sum, from a Marxist perspective the north-Pacific seal-herd case demonstrates the systematic (rather than accidental) tendency of capitalist societies to overexploit their natural-resource base. This overexploitation is a product of the *articulation* of a particular mode of production with the specific physical capacities of resources and environment. These capacities, while quite real and tangible, are not absolute though. Rather, they are seen as being relative to the demands made on them by the mode of production.

The production of nature

The society–environment dialectic examined in the previous subsection is, if you like, an 'external' one. It involves situations where societies are confronted with natural phenomena that are not readily amenable to physical manipulation. These phenomena can be destroyed by societies but not created or controlled by them. Here, as Boyd et al. (2001: 557) put it, actors in capitalist societies 'confront nature as an exogenous set of material properties'. This is especially true in extractive sectors of the economy like mining, fishing, sealing and whaling (for instance, diamonds are given in nature: they can be mimicked in laboratories but not made like for like). However, things are different in farming and forestry – two cultivation-based sectors of the economy. In these sectors, ways have been found to produce nature 'all the way down': that is, physically. The term 'the production of nature' was coined by Neil Smith in 1984 – over a decade before genetically modified crops, transgenic animals and bioengineered trees made it seem a prescient rather than fanciful one. As a Marxist, Smith insisted that firms operating in a capitalist economy will seek to overcome the 'barriers to accumulation' that are thrown up by the non-human world. For him, these firms will try to find ways to 'making nature to order' in order to realise profits. Here, the non-human world becomes a mere *means* to the end of profit-making: the overriding objective of firms in capitalist societies. For Smith, then, nature is becoming increasingly 'internal' to the logic of capitalist societies. It is a 'second nature' far removed from the 'first nature' bequeathed by evolution.

Smith's point may now seem rather obvious given that such things as genetically modified organisms are today a recognised part of the product range of biotechnology firms (like Monsanto). However, from his perspective it is still important to understand precisely *how* and *with what consequences* (social and ecological) particular elements of the non-human world are being 'produced' (see Smith 1996). What's more, when Smith was writing back in 1984 he was not predicting the future, but, rather, talking about the present and the past. In other words, he was arguing that elements of the non-human world had been materially produced for a very long time. For him, the common-sense belief that society and environment are two separate physical domains was blinding people to what was going on under their very noses: namely, the deliberate alteration of non-human entities by a selection of capitalist enterprises for profit purposes, rather than any higher goals. The Marxist sociologist Jack Kloppenburg (1988) has provided a now-classic analysis of the production of nature in the era preceding our own. His book *First the Seed* explained how and why capitalist firms gained physical and proprietary control over one of the biological bases of commercial agriculture world-wide: namely, seeds. This analysis of seeds, which I will now summarise, exemplified empirically what Neil Smith argued theoretically: that capitalist firms increasingly fabricate a non-natural nature with potentially dire human and environmental consequences.

Seeds are 'the irreducible core of crop production' (Kloppenburg 1988: xi). Farmers need them for virtually all crops and, for this reason, they are an attractive profit possibility for capitalist firms. However, historically, there has been a natural barrier to a capitalist seed industry developing: namely, the fact that seeds are self-reproducing. In nature, each new harvest yields seeds for next year's crop, meaning that farmers have enjoyed a ready-made – and free – supply of seeds for millennia. As the Marxist sociologists Mann and Dickinson (1978: 467) aptly put it, 'Capitalist development appears to stop, as it were, at the farm gates'. However, from the early twentieth century, this changed. As Kloppenburg shows in fascinating detail, the natural barrier that seeds presented to the development of an off-farm seed industry was overcome. Let's take the case of corn (maize), a crop that has long been of immense importance within modern agriculture. Unlike other major crops, corn is naturally open- or cross-pollinated. In contrast to self-pollinated crops (such as wheat), corn plants are the product of a unique mix such that a field of corn is 'in a constant state of genetic flux' (1988: 95). Clearly, this natural promiscuity poses a major natural barrier for potential crop breeders: for 'superior' corn plants with desirable characteristics (e.g. the capacity to withstand disease) are constant admixing with 'inferior' corn plants.

So how was corn reproduction controlled? And what were the consequences? Focusing on the USA, Kloppenburg's answer to the first question is that the discoveries of geneticist Gregor Mendel fortuitously opened the door for the development of hybrid corn: that is, corn bred off-farm with superior characteristics. Mendel's late nineteenth-century experiments showed that it was possible to cross-breed strains of a crop in a controlled, systematic way. In the 1890s the US government had created a system of land-grant universities and agricultural research stations devoted to helping the nation's farmers improve the quantity and quality of crop yields. Scientists employed in these universities and stations were soon able to alter the character of most major crops, including corn. As one of these scientists put it, 'The . . . breeder's new conception of [crop] varieties as *plastic groups* must replace the old idea of fixed forms of chance origin which has long been a bar to progress' (W.A. Orton cited in Kloppenberg 1988: 69). Using complex, time-consuming procedures, plant-breeders like Orton were able to produce very high-yielding strains of corn (see Figure 3.1). Aside from the complexity of producing the seeds for this 'double-cross' corn, and despite its high-yielding character, it was also 'naturally sterile'. In other words, the seed from double-cross corn turned out to produce low-yielding crops if planted, forcing farmers to go back to the universities and research stations who provided the seed for the improved corn the year before. By accident, then, rather than by design, a potential new market in corn seed was produced in early twentieth-century America. As Kloppenburg shows for corn as well as other hybrid crops, the commercial implications of this were quickly recognised. Several government plant-breeders sought private investment, resigned from their posts, and set about producing improved seeds for sale to farmers in newly established seed companies (see Figure 3.2). By the 1930s, having established biological control over the reproduction of most major crops in the USA, these firms recognised the need for legal control of their seed 'inventions'. The 1930 Plant Variety Protection Act was the result of heavy lobbying by private seed firms in the USA. It entitled these firms to full legal control of their seeds so that rival producers could not copy these seeds without financial compensation.

In sum, Kloppenburg's book shows how and why nature (in this case, seeds) was materially produced as part of conscious accumulation strategy among capitalist firms. More than merely 'tampering with' or 'disturbing' the biophysical functioning of crops, the seed companies whose activities Kloppenburg investigates actively reconstituted the 'nature' of those crops. The consequences of the production of seeds make for an interesting comparison with the current furore in some parts of the world over genetically modified crops. On the environmental side, Kloppenburg showed how hybrid varieties led to monocultures becoming dominant in US agriculture. Monocultures are croplands of a genetically uniform nature. While high-yielding, these croplands require heavy doses of herbicides and pesticides in order to protect them against weeds and pests. The environmental knock-on of this has been decades of polluted soils and watercourses (famously identified by Rachel Carson in *Silent Spring*). These knock-on effects have been the 'unintended consequences' of intentional productions of agrarian nature. Socially, Kloppenburg's analysis pointed to the disenfranchising effects of 'outdoing' nature in commercial agriculture. For nearly a century, farmers in the USA and beyond have lost their right to a previously free, 'public domain' good. They must now pay often-high prices not just for seeds but also the chemical treatments that must be applied to engineered crops in order to ensure their healthy growth. From the perspective of Marxists like Smith and Kloppenburg, only time will tell

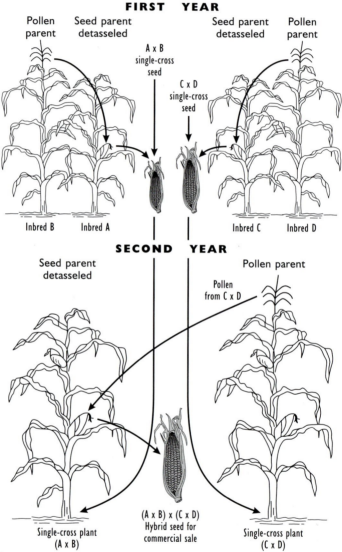

FIRST YEAR

Pollen parent | Seed parent detasseled

A x B single-cross seed

C x D single-cross seed

Seed parent detasseled | Pollen parent

Inbred B Inbred A

Inbred C Inbred D

SECOND YEAR

Seed parent detasseled

Pollen parent

Pollen from C x D

(A x B) x (C x D) Hybrid seed for commercial sale

Single-cross plant (A x B)

Single-cross plant (C x D)

Seed Company

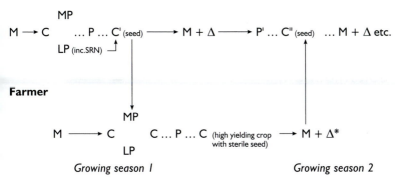

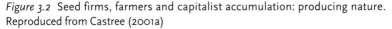

Growing season 1   *Growing season 2*

Figure 3.2 Seed firms, farmers and capitalist accumulation: producing nature. Reproduced from Castree (2001a)

whether genetically modified crops exacerbate the environmental and social ills associated with hybrid crops.

It is, I hope, clear from the discussion above that the arguments of Smith, Kloppenburg and other like-minded Marxists do not amount to the absurd claim that capitalist firms have complete *control* of the non-human world. For these authors significant parts of that world are indeed 'constructed' in the most physical of ways. But this is not the same as saying that produced nature lacks a certain agency and unpredictability of its own. For instance, the biotechnology firms who have vigorously promoted GM crops for over a decade admit that they cannot anticipate the impact these crops will have on other flora and fauna. Whatever the impact may be, from the

---

Figure 3.1 (opposite) Production of double-cross hybrid seed corn using manual detasseling. Reproduced from Kloppenburg (1988: 100)

Note: The process begins with two pairs of homozygous inbred lines (A, B and C, D). Each pair is crossed (A × B, C × D) by planting the two lines in alternating rows and emasculating the femal parent by manual removal of the pollen-shedding tassel (this process is known as detasseling). Only seed from the female parents is collected top ensure that no selfed seed is obtained. Plants grown from this single-cross seed are again crossed following the same procedure: (A × B) × (C × D). seed is again collected from the female parent, and it is this germplasm that is the double-cross hybrid seed sold for farm production

production-of-nature perspective it is part of an internal dialectic: one contained *within* the capitalist system rather than *between* it and an unproduced nature.

WHY ARGUE THAT NATURE IS A SOCIAL CONSTRUCTION?

In this chapter I have explored the two main strands of constructionist argumentation in contemporary human and environmental geography vis-à-vis nature. I have dwelt on the representational strand at greater length than the material strand because it has been so dominant in human and environmental geographers' analyses of nature this past decade or so. Together, these strands have emphatically de-naturalised our understanding of what nature is, how it works, how we should evaluate it, and how we should use it. As I will explain in some detail in Chapter 4, the de-naturalising approach contrasts strongly with how physical geographers understand nature. Outside geography specifically, and academia more generally, many people find the suggestion that nature is not natural absurd or scandalous – like the environmentalists who objected so strongly to William Cronon's argument that wilderness is an idea not a reality. In this penultimate section of the chapter, I discuss what motivates so many contemporary geographers to argue that what we call nature is a social construction. In keeping with my overall approach to nature in this book, I explore what is to be gained (and lost) when students, academics and groups outside the university are persuaded that nature is not what it seems to be (i.e. 'natural'). Note that I deliberately do not discuss whether social-constructionist authors are right or wrong in the claims they make. Others have debated this issue vigorously and I leave it to readers to make up their own minds (see Box 3.5).

ACTIVITY 3.6

In your view, what motivates geographers whose research shows that nature is a social construction? A different way of phrasing this question is to ask: what problems do these geographers see attaching to the idea that nature is natural?

Box 3.5 IS NATURE A SOCIAL CONSTRUCTION?

The debate over whether or not nature is a social construction extends way beyond the discipline of geography. Sociologists of environment and the body, environmental historians, sociologists and anthropologists of science, environmental anthropologists and philosophers (among others) have all recently jousted over the issue. The debate is a complex and often heated one. Any reasoned answer to the question 'Is nature a social construction?' would have to carefully identify which 'nature' is being discussed, what kind of 'social constructionism' is being discussed (representational or material) and what degree of constructionism is at issue ('strong' or 'mild': see Demeritt 2002). In geography, the debate about the social construction of nature has focused mainly on the environment. (See Proctor 1998; 2001; Demeritt 2001b; Peterson 1999 and several essays in Cronon 1996. Outside geography, the essays by Lease, Shepard, Hayles and Soule in Soule and Lease 1995 are useful, as is the collection edited by Bennett and Chaloupka 1993; see also Bird 1987; Burningham and Cooper 1999; Gifford 1996 and Greider and Garkovich 1994). However, it has also spiralled out into a debate over the social construction of reality (or knowledge about that reality). Here 'realists' have confronted 'relativists'. One of the best examples of this confrontation is that between the transcendental realist Andrew Sayer and human geographers of a more postmodern and post-structural persuasion (see Sayer 1993 and the references therein to Strohmayer and Hannah). Within particular branches of geography the debate over nature's social construction has also been evident. For instance, many geographers interested in the disabled have questioned whether 'disability' is a bodily and mental fact as opposed to a social construction of certain peoples' bodily and mental condition (see Butler and Parr 1999, Kitchin 2000 and Imrie 1996). Likewise, in recent years rural geographers have questioned the view that the countryside is more inherently natural than towns and cities (see Bunce, 2003 and the references cited therein; see also Castree and Braun 2005). For general introductions to social constructionism from a social-science perspective see Gergen (2001) and Burr (1995). Within social science, three of the main areas where the social construction of (human) nature has been hotly debated are the areas of identity, the body and 'race'. The last of these has been a highly contentious topic for the obvious reason that racial categorisations – which frequently rest on claims that people can be distinguished according to their mental and physical characteristics – have long been used to discriminate against certain people. For more on the debate over 'race', nature and social constructionism within and beyond geography see Duncan et al. (2004: ch. 16), Wade (2002), Miles (1989), Malik (1996) and Banton (1998).

A useful way to answer the Activity question is to distinguish between cognitive, moral and practical reasons. You will recall from Chapter 1 that cognitive claims are descriptive and/or explanatory claims about real-world phenomena (or existing ideas about these phenomena). Moral (or ethical) claims, meanwhile, are value-judgements about particular things or ideas and they can be *normative*, specifying how we *should* value those things or ideas. Depending on how we view (cognise) and value things, certain practical consequences follow. Using this three-part schema, we can understand the appeal of social-constructionists' arguments for those who advocate them. Cognitively, the geographers whose work I've examined in this chapter – despite their different theoretical and empirical emphases – have one key thing in common. Each of them insist that when we (as geographers and ordinary people) talk about or attribute something to 'nature' (in any of the three main senses of the term) we are often making a *category mistake*. For them, we are confusing representations with the realities they apparently reflect or else failing to see that those realities are no longer 'natural' (in the sense of untouched by society). Consequently, it is not just de-mythologising research (like Forsyth's) or ideology-critique arguments (like Harvey's) that hold fast to the conviction that there is a 'right' and 'wrong' way to understand what nature is. All the geographical research discussed in this chapter maintains, at some level, that it is simply incorrect to suppose that nature is natural.

The moral implications of this cognitive move are considerable. First, they take us away from moral naturalism, which is the belief that nature offers us moral lessons that we ought to obey (see Box 3.3). Moral naturalism can be very authoritarian. It can close down discussion about appropriate moral behaviour by trying to 'read off' an ethical code from the supposed 'facts of nature'. From the perspective of geographers who see nature as a social construction in degree or kind, moral naturalism is disingenuous. It conceals a situation where one or other person's morality is being surreptitiously promoted by attributing that morality to a realm that is either non-human or else a part of 'human nature' that is supposedly unalterable and given. For instance, it was once common in Western countries to hear the refrain that women are 'naturally' caring, emotional and nurturing, while men are naturally rational and analytical. The moral lesson drawn from these two natural 'facts' about men and women was that the latter should stay at home to raise children, while the former should be breadwinners. With the benefit of hindsight, we would now say that this morality does not reflect a natural state of affairs but, rather, the norms of patriarchal society where men were (and arguably still are) highly focused on controlling the life-chances of women.

Second, the critique of moral naturalism implies that we should be more honest about the social constructedness of *all* moral values. For social constructivists, a non-naturalistic morality is quite liberating. It shows us that what we deem to be morally good or bad, right or wrong, fair and unfair, is a societal *choice*, not something dictated to us by a putatively non-social nature. This means that those who, for whatever reason, oppose dominant moral norms, should feel empowered to challenge them: to make a case for a different morality rather than appeal to some non-social realm as a source of values. Third, this does not mean that social constructivist geographers do not care about those things that societies normally think of as 'natural' (like otters or trees or the ozone layer). All it means is that most of these geographers see moral values as deriving from society not nature. Thus, one can still be an ecocentrist adhering to a 'green' morality while being a representational or material constructivist. But one would recognise that this moral stance is not mandated by the non-human world but by a set of contingent values that happen to see that world as precious and important. This said, it must be acknowledged that few geographers have set about developing an ecocentric morality, from within or without the social-constructivist camp. I suspect that the reaction to moral naturalism has been so strong among left-wing geographers that they have felt disinclined to moralise about nature, even within a social-constructivist framework. Overall, geography's 'social' and 'cultural left' has outgunned its barely existing 'green left' – notwithstanding the efforts of some (like Gavin Bridge) to take seriously the environmental impacts of human action.

In the fourth place, those social-constructivist geographers who are preoccupied with people's identity and corporeality (rather than 'external nature') tend to be moral pluralists. Their critique of moral naturalism is designed to valorise identities and bodily practices that have been stigmatised by their classification as 'unnatural'. For instance, critical geographers studying the physically and mentally disabled have advocated the so-called 'social' explanation of disability over the 'medical' one. They have shown that the stigmatisation of disabled people is not simply a result of their objective differences from 'able'-minded and -bodied individuals. Rather, it is also the result of the socially entrenched attitudes used to make sense of disabilities. Thus, by challenging the idea that disabled people's role in society is prescribed by their physical condition, many critical geographers have created a moral space where the needs, rights and entitlements of the disabled can be placed on a par with those of everyone else in society (see Imrie 1996). Taking this further, several other left-wing geographers have used the denaturalisation of subjectivity and corporeality to demonstrate the diversity of moral agendas within seemingly unified groups (like women or people of colour). For instance, 'second-wave' feminist geographers have questioned the idea that women's biology and psychology is similar or universally shared by accenting the diverse ways that people classified as women think and act. These feminists show how women are interpellated within multiple discourses (of gender, 'race', class etc.) such that their moral values and aspirations vary a great deal (Pratt and Hanson 1994). In this sense, their research is 'anti-essentialist': it maintains that 'what essentialists "naturalize" . . . is actually socially constructed difference' (Valentine 2001: 19).

Finally, the practical implications of social constructivists' arguments are potentially profound for those phenomena that constructivists wish to de-naturalise. In Chapter 1 we learnt that one of the principal meanings of the term nature is 'the essence of something'. If a person's mind or body, or the non-human world, has an essence then it would appear to be a relatively fixed and unalterable aspect of the phenomenon in question.

However, as we've seen in this chapter, critical human and environmental geographers have shown that claims are made about the essence of the non-human world or the human mind and body are rarely innocent. For instance, the practical actions that follow from the neo-Malthusian reasoning criticised by David Harvey in 1974 are population control or else *inaction* (i.e. let those who 'breed too much' starve). For left-wing geographers such policies are oppressive and callous. Likewise, when representations of black men as 'inherently athletic' circulate in society they can have pernicious effects. As Peter Jackson's (1994) analysis of Lucozade ads starring Olympic decathlete Daley Thompson showed, these representations can tacitly support policies where black men are offered little support in their quest for mental as opposed to manual occupations (see Lewis 2001). The flip side of this critique of essentialist claims is that constructivist geographers have highlighted the potential *alterability* of many things in our world which appear unchangeable. As Demeritt (2002: 769) puts it, '[One] . . . objective of denaturalisation is to show that something is bad and that we would be better off if it were radically changed, which becomes conceivable once we realize it is socially constructed and within our power to change'.

The cognitive, moral and practical intentions behind social constructionist arguments are not above criticism (see Box 3.6). But they nonetheless amount to a powerful and distinctive stance on what nature is, how it is (or should) be valued, and what to do to those things routinely designated as 'natural'. Where many in academia and the wider society predicate their views about nature on the assumption that nature is separate from or different to society, social-constructivist geographers adamantly resist this separation. These geographers have little time for ontological, causal, moral or other references to an asocial nature. For them such references are misguided, if not downright dishonest and dissimulating (see Box 3.7).

SUMMARY

This chapter has surveyed de-naturalising approaches to nature in contemporary geography. After a discussion of the precursors to these approaches, the chapter explained the representational and material variants of social constructionism and subvarieties thereof. This led to a discussion of the reasons why so many present-day geographers are keen to produce de-naturalising understandings of those phenomena normally thought to

Box 3.6 THE SOCIAL CONSTRUCTION OF NATURE: THREE SHIBBOLETHS

Left-wing human and environmental geographers arguably adhere to some shibboleths about that which we call 'nature'. Shibboleths are beliefs that like-minded people take for granted as true, correct or otherwise accurate: they are shared assumptions that characterise the views of a given group. Among social-constructionist geographers three shibboleths about nature loom large. The first two relate to the way nature is depicted within socially dominant representations of the non-human world and the human mind or body. The first shibboleth is that these representations are typically about the fixity and permanence of nature. We've seen this several times in this chapter. Geographers from Harvey to Braun criticise references to the moral or practical 'lessons' that nature supposedly teaches us by virtue of its 'essential character'. Such criticisms are valid as far as they go. There is no doubt that many nefarious beliefs and practices have been justified by self-serving references to 'natural imperatives'. For instance, in both the past and present, racism and sexism have gained legitimacy by instilling the belief that there are 'natural differences' between people that are unalterable. However, as Wade (2002) points out, references to the 'nature' of the non-human world or human psychology and physiology need not *necessarily* imply permanence and fixity. For instance, in the present period there is veritable obsession within sections of Western society with the conscious *altering* of the human body. Such practices as dieting, body-building and plastic surgery all rest on the idea that a person's physiology is changeable and not simply an unavoidable fact of bodily nature. Fitness clubs, health-food firms and biomedical companies all have a vested interest in promoting these practices rather than in emphasising the givenness of the body. The second social-constructionist shibboleth we can question is that most references to the fixity of nature serve the moral and practical interests of the powerful. Though this shibboleth has much validity, it ignores the fact that many individuals and groups who suffer marginalisation or oppression can use claims about the supposed intransigence of their bodies or minds to their advantage. For instance, suppose one argued that the sexual preferences of gay and lesbian individuals is genetically caused – something a person inherits at birth. This argument can be used to suggest to homophobic people that homosexuality is every bit as 'normal' a part of human biology as is heterosexuality. The third and final social-constructionist shibboleth we can question relates to the practical implications of de-naturalising arguments. As the section 'Why argue that nature is a social construction?' on pp. 166–71 explains, social-constructionist geographers emphasise the potential malleability of that which appears to be fixed and natural. However, it is unclear why showing something to be 'social' makes it any more amenable to change or improvement. After all, social power, social relations and social attitudes are often as difficult to alter as human DNA or the orbit of the moon. For instance, if one follows Foucauldian reasoning, then dominant discourses are extremely hard to change because they are so deeply embedded in people's thoughts and actions.

Box 3.7 HUMAN GEOGRAPHY, THE SOCIAL SCIENCES AND THE 'NEW NATURALISM'

Over the past ten years or so, nature has been brought 'back in' to several social-science disciplines in a far more literal (i.e. non-social constructionist) way than it has in human geography. In sociology and anthropology, for instance, there have been attempts to marry natural-science insights about human physiology, human psychology and the non-human world with social-science insights into why people think and act in the ways they do. In its strongest form, this marriage seeks to explain social phenomena with direct reference to natural phenomena. 'Socio-biology', for example, traces links between people's genetic make-up and their characteristic modes of behaving. In its crudest form, socio-biology is biologically determinist and has found expression is racist tomes such as Hernstein and Murray's (1996) *The Bell Curve*. However, it would be a mistake to think that all attempts to explain social phenomena in terms of human and non-human nature are reactionary and conservative. The radical sociologist Ted Benton (1994), for example, argues that the biological needs of human beings can offer us a reference point for criticising the harm that we do to the non-human world and our own bodies in advanced capitalist societies. For introductory, and sometimes critical, discussions of the new naturalism in the social sciences see Barry (1999: ch. 8), Benton (1994), Dickens (2000) and Ross (1994: ch. 5).

be natural. I have refrained from offering an assessment of whether 'nature' really is a social construction. Though this is an important issue that readers should consider (see Box 3.5), my aim has been to focus on how and why many geographers produce knowledges that are 'nature-sceptical'. In the next chapter, I want to examine a very different set of knowledges about nature. They are knowledges of the non-human world (rather than the human body) but ones unlike those presented in this chapter. These knowledges are produced by physical (and many environmental) geographers. These geographers, as we'll now discover, argue that their representations of nature can and do represent a natural world that is irreducible to people's ideas or their practices. By denying that nature is (or is only) a social construction, these geographers are trying to persuade people within and beyond universities that their knowledge is preferable to that promulgated by those whose work I've discussed in this chapter. Here, then, we have a contest – one almost entirely implicit than explicit it must be said – between knowledges of nature: a tussle over whose knowledge is the most accurate and appropriate.

EXERCISES

- List some of the problems that arise if one accepts the impossibility of ever talking about nature 'as it really is'. To start you off, one problem is that if we can never confidently identify environmental problems when they arise then we may be slow to act and may suffer the consequences later.

- Do you think that factual claims about what is (or is not) 'natural' can be disentangled from moral claims about nature? Consider the issue of 'race', for instance. Racial differences between people are often thought to be biological differences (of genotype and/or phenotype). If, as social constructionism argues, racial differences are not given in nature then what, if anything, follows morally from demonstrating this? Does it *automatically* follow that we should condemn or criticise people who hold on to naturalistic views of racial difference?

FURTHER READING

Obviously, readers should consult the works by Harvey, Watts, Forsyth, Moore, Cronon, Braun, Bartram and Shobrook, Castree, and Kloppenburg (1988: chs 1 and 2) discussed in this chapter. These are best read immediately after the section of the chapter in which they are summarised. For further information on the social construction of nature debate see the readings cited in Box 3.5. The follow-up readings for the various parts of this chapter are extensive and relate to its main sections on the discursive and material construction of nature. Some of these readings are also cited in Box 3.5.

For more on the relationship between population and resources see Woods (1986), Bradley (1986), Findlay, (1995), Halfon (1997), Maclaughlin (1999), Norton (2000), Petrucci (2000), and Taylor and Garcia-Barrios (1999). Natural-hazards research in the post-Hewitt period is well discussed by Abramovitz (2001), Blaikie et al. (1994) and Pelling (2001; 2003).

There is now a tremendous volume of research by geographers into representations of nature. Often this research mixes and matches the four approaches to representation I've identified in this chapter (plus some others). Geography journals that have routinely published this research for a decade are *Environment and Planning D: Society and Space, Geoforum* and *Ecumene* (now called *Cultural Geographies*). For more on post-structural and Foucauldian theories of the environment see Conley (1997) and Darier (1999). For more on Marxist ideas about the society–environment dialectic ('internal' and 'external') as well as the material 'production of nature' see Castree (2000; 2001a) and Boyd et al. (2001).

Two subfields of geography where the representation of nature and the materiality of nature respectively have been important are Third World political ecology and animal geography. In the first case concerns have been expressed that a focus on discourses of environment has undermined understanding of the real biophysical world. In the latter case, the agency of animals has been emphasised as a counter to the tendency of human geographers to ignore the physicality of the non-human domain. For more on Third World political ecology see Peet and Watts (1996), Robbins (2004b) and Zimmerer and Bassett (2003). For an introduction to animal geography see Wolch and Emel (1998) and Philo and Wilbert (2000).

There is no one text on how human geographers have de-naturalised our understanding of people's identities and subjectivities. However, Panelli's (2004) book is very good indeed (though not structured explicitly around the theme of de-naturalisation), while chapters 7–8 and 16–18 of Duncan et al. (2004) are also most useful. More generally, Barker (2000: ch. 6) gives a useful overview of de-naturalising approaches to understanding subjectivity and identity. Kay Anderson, in her work on 'race' (2001), has been a leading analyst of how conceptions of mental and biological essentialism are used to justify discrimination; see also Penrose (2003).

I argued in this chapter and the previous one that geographers had said little about an ethics of nature – whether in a naturalistic or de-naturalising mode. The following are among the few by human geographers on ethics and nature: Low (1999), Lynn (1998), Jones (2000), Low and Gleeson (1998: ch. 6), Proctor and Smith (1998: section III). The annual 'progress reports' on ethics in *Progress in Human Geography* mention new literature on nature ethics by human geographers. Proctor (2001) discusses whether it is possible to have a nature-sceptical ethics of nature, as does Petersen (1999). These two essays are essential reading for understanding whether a 'social-constructionst ethics' of nature is viable or desirable.

4

TWO NATURES?
The dis/unity of geography

'We part company with [other geographers] by acknowledging the possibility of identifying and studying "real" . . . processes'.
(Slaymaker and Spencer 1998: 248)

'The naturalness of nature is, in one sense, inherently self-evident'.
(Adams 1996: 82)

INTRODUCTION

The first epigraph, taken from *Physical Geography and Global Environmental Change*,
is a sideswipe at the de-naturalising approaches to nature explored in the
previous chapter. Like virtually all physical geographers, Slaymaker and
Spencer see themselves as scientists: people who are in the business of
producing accurate knowledge about the workings of the non-human
world (they leave investigations of the human body to others within and
beyond geography). While these two leading geomorphologists do not
deny that what we call nature is often at some level 'unnatural', they
nonetheless maintain that it has distinct ways of working that need to be
comprehended objectively. In other words, they see it as irreducible to
particular social representations and practices and as amenable to relatively

unbiased analysis. Likewise, the environmental geographer Bill Adams takes it as axiomatic that nature is, wholly or in part, 'natural'. Adams's (1996) book *Future Nature* reflects the views of many in geography's 'middle ground'. It argues that geographers should study human uses and abuses of the environment so as to fashion more effective conservation and restoration policies – effective because they are based on accurate understanding.

The suspicion about social constructionism evinced in the quotes from Slaymaker, Spencer and Adams seems intuitively legitimate for several reasons. First, one can argue that many aspects of the non-human world (and, for that matter, the human mind and body) exist regardless of how we represent them – like continental plates or Mount Everest. Likewise, even though we actively fashion our knowledge of nature this does not necessarily make it false, inaccurate or untrue. As Slaymaker and Spencer argue, our representations of nature may be constructed but they may also be *accurate constructions* if arrived at using appropriate procedures. Third, while Marxist geographers (among others) may be right that societies can physically produce some parts of what we call nature 'all the way down', even these produced parts arguably have a 'nature' that is irreducible to the social processes that gave rise to them in the first place. Finally, we can challenge the metaphor of 'construction' when applied to many aspects of nature. Take acid deposition, for example. This environmental problem has undoubtedly been created by human action. But it's a phenomenon that is an unintentional consequence of our activities and which has a certain life of its own (by virtue of our inability to control the atmospheric dynamics transporting various oxide pollutants across oceans and land masses). Acid deposition is thus, perhaps, best thought of as a 'manu-factured risk' (Beck 1992) rather than a 'social construction'. In the eyes of some geographers, the latter term implies a degree of intentionality and control that is absent in this and many other cases where people alter the environment.

For these four reasons (and others not mentioned), physical geographers and many environmental geographers are more 'nature-endorsing' than their social-constructionist counterparts. They take it for granted that (i) the non-human world exists independently of our representations of and actions upon it, and (ii) that, in both principle and practice, it can be understood in more or less accurate ways. In this chapter I want to explore why and how physical geographers adhere to this belief in the reality of the non-human world and its capacity to be understood more or less objectively. I focus on the physical side of geography because it wears its realist credentials on its shirtsleeves. By 'realist' I mean beliefs (i) and (ii) above, the first being so-called ontological realism, the second being so-called epistemological realism.[1] This links closely with physical geography's self-image as a science. It's fair to say that most physical geographers regard themselves as scientists. As Clifford (2001: 387) confidently asserts, 'the first presupposition is that physical geography is . . . a scientific activity'. Indeed, physical geography is about the only part of geography where the word 'science' is still used openly and unself-consciously to characterise the conduct of research. Since the word is usually associated with the search for truth and objectivity about the material world it follows that physical geography eschews the apparent anti-realism that Slaymaker and Spencer associate with social-constructionist approaches to nature. It also follows that physical geographers aim to produce cognitive knowledge for the most part, taking it as read that statements of fact and statements of value (moral and aesthetic) should be kept separate. In effect, the naturalism of physical geography is the mirror opposite of the de-naturalising thrust of contemporary human geography (leaving environmental geography a schizophrenic field with, as it were, 'divided loyalties'). As Turner (2002: 62) puts it, 'With physical geography esconced in the sciences and much of human geography engaged in various experiments that challenge this way of knowing, the gulf between the two appears to have widened'.

In the next section I explore how physical geographers characterise their half of geography and how, in broad terms, they might defend their quest for accurate knowledge about the non-human world. I then question physical geography's epistemically realist credentials by exploring the idea that even scientific knowledge of nature is a social construction. This leads to a discussion of how physical geographers have sidestepped the social-constructionist critique of the knowledge they produce. I then show that the key debates in physical geography revolve around producing not accurate knowledge of the biophysical world (since this possibility is largely taken for granted) but *more accurate* knowledge of that world.[2] I conclude by reflecting on how the co-existence of constructivist and realist approaches to nature within geography is central to the ongoing estrangement of human and physical geography. Before I proceed I should declare a crucial gap in my discussion: because I treat physical geography as a field science, I inevitably ignore important non-field based activities, like numerical and computer modelling.

ENVIRONMENTAL REALISM: AGENDAS AND JUSTIFICATIONS

Physical geographers rarely reflect, in any formal way, on what makes their field of study a 'science'. Yet it is surely no accident that they use the appellation to describe their research. Science is a highly loaded word. The key to its power is that it's uniquely associated with the ideals of truth, objectivity and accuracy (Box 4.1). As Alan Chalmers (1999: 1) put it in his book *What Is This Thing Called Science?*, 'scientific knowledge is [seen as] proven knowledge'. This echoes the view of the influential philosopher of science Karl Popper: 'science is one of the few human activities – perhaps the only one – in which errors are systematically criticised and fairly often, in time, corrected. In most other fields of human endeavour there is change, but rarely progress' (Popper 1974: 216–17). The word 'science' does not simply describe a set of investigative procedures that anyone who wishes to be a scientist should adopt if they are to produce accurate knowledge about a given phenomena. More pointedly, it is also a rhetorical weapon. As I explained in Chapter 2, geography as a whole began to self-consciously characterise itself as a science from the 1950s in response to pressures emanating from outside the discipline and as a means of effecting intellectual change within the discipline. Though the term had been used to describe geography since the discipline's inception, it took on a more *substantive* meaning from the mid-twentieth century. I will say more about that substantive meaning later. For now I simply note that the appellation 'science' served political as much as intellectual purposes (Castree 2004a). In the case of physical geography, not only did it permit criticism of previously dominant research approaches (like W. M. Davis's rather speculative ideas about landform evolution). It also allowed physical geographers to align their research with 'prestige' disciplines like physics, chemistry and biology and so boost their image within and outside geography.

More than five decades on, most physical geographers describe themselves as scientists as a matter of course. The scientific status of their research is simply taken for granted. This might suggest that physical geographers actively and frequently discuss what is 'scientific' about their mode of interrogating the world. But in reality such discussion is rare. Both prior to and between texts like *Physical Geography: Its Nature and Methods* (Haines-Young and Petch 1986) and *The Scientific Nature of Geomorphology* (Rhoads and Thorn 1996) there were few formal discussions of science by physical

Box 4.1 SCIENCE AND PHYSICAL GEOGRAPHY

There is no one definition of science. Those offered by historians and philosophers of science tend to be either *positive* or *normative*. Positive views of science define it with reference to how people who call themselves 'scientists' actually undertake research. Normative views, by contrast, lay out a template for how researchers *should* investigate the world if they're to qualify as being scientists. Simplifying, we can say that any full definition of what science is (or how it should be practised) should make reference to three things: namely, a set of axiomatic beliefs ('the scientific worldview'), an investigative procedure ('the scientific method') and a product that emerges from these two things (scientific knowledge). In physical geography some basic beliefs that are taken to be axiomatic are as follows (clearly, these will vary from researcher to researcher): (i) the non-human world is real and its characteristics are irreducible to any given set of human perceptions about, or practices upon, it (this belief is sometimes called materialism); (ii) the non-human world has an inherent order which, however complex, is amenable to discovery; and (iii) though we may value the non-human world in moral and aesthetic ways, science is concerned primarily with cognitive matters (e.g. matters of fact, explanation and prediction). More generally, many physical geographers would be comfortable with the third and fourth 'scientific norms' identified by Robert Merton back in 1942: namely, that science is disinterested (free from prejudice) and that it is organised scepticism (it only accepts statements about the world if they can be proven to be true). On the basis of these broad, shared assumptions, physical geographers have an equally broad commitment to a mode of interrogating reality that, in their view, can produce knowledge that accurately captures its truths. Though there is no single scientific method employed by physical geographers, Schumm (1991) is probably right that there are some general investigative steps that most practitioners adhere to. These will be discussed in the section 'Producing realistic environmental knowledge' on pp. 191–202 of this chapter. Finally,

> it is the rigour with which these steps are followed that gives physical geographers confidence that the knowledge they produce is realistic rather than false. As Schumm (1991: 26) put it, 'method is only as powerful as the objectivity of the individual using it'.

geographers, and there have been fewer still since. In the main, physical geographers prefer to 'do' rather than to philosophise about the manner of their doings. For them, the scientific nature of their research is manifest precisely in its execution. In other words, physical geographers have not posited an ideal model of Science (with a capital S) to which their research should conform. Though they have drawn inspiration from philosophers and historians of science, they have not mechanically adhered to received notions of how 'proper scientific research' should proceed.

The reason for this, in part, is that many of these notions have been derived from laboratory science. Yet most physical geographers would characterise themselves as field scientists (Phillips 1999: 482). Unlike laboratory scientists, field scientists investigate the non-human world in 'live' rather than 'artificial' settings.[3] Field sciences are typically 'composite' disciplines whose aim is synthesis. They combine knowledge from other sciences and apply it to an understanding of complex and often dynamic environments that are not readily amenable to experimental control. Thus, physical geography draws upon physics, chemistry, mathematics and biology to aid its understanding of biophysical reality. But this does not make it a purely derivative discipline, reliant on others for knowledge and understanding. The distinctiveness and originality of physical geography is that it seeks to understand how the phenomena studied in relative isolation by other natural sciences *come together* in specific spatio-temporal contexts. As Ken Gregory (2000: 9) put it in his well-known definition: 'Physical geography focuses upon the character of, and processes shaping, the land-surface of the earth and its envelope, emphasizing the spatial variations . . . and temporal changes necessary to understand the . . . environments of the earth . . .'. Like earth science and environmental science, physical geography 'is concerned with phenomena with many interacting parts' (Malanson 1999: 747). For instance, a fluvial geomorphologist studying gravel-bed rivers needs to understand the relationships between (i) water volume, speed and turbulence, (ii) the character of bed gravel (iii) the nature of aquatic flora and fauna (iv) sediment load, and (v) the erosivity of river-bank material (among other things). Because rivers are 'open systems', these relationships cannot be studied in the same way as a laboratory scientist studies a 'closed system' – one where the variables of interest can be isolated and held constant. In sum, and to simplify somewhat, physical geography's niche within the sciences lies in its aspiration to produce accurate knowledge of the interactions that give the non-human world its particular character at particular spatial and temporal scales.[4]

In light of the above discussion, it's clear that geography is a divided discipline when it comes to knowledges of nature. As explained in the previous chapter, those human geographers who study nature are concerned with the ways things so named are understood and materially altered by people. Though they claim that their knowledge of what we call nature is accurate (while being sceptical of other people's), they mostly prefer not to characterise their methods or their findings as 'scientific'. The reasons for them rejecting the appellation 'science' are complex, but among them is the fact that human geography is currently too intellectually diverse for this label to serve as an adequate descriptor (see Demeritt 1996: 486–90). By contrast, physical geographers are interested in the non-human world in and of itself – rather than in how that world is understood by society or in the social practices and forces altering the character of that world. As Urban and Rhoads (2003: 224) express it, 'The domain of physical geography is the biophysical world. If humans are considered it is only the effect of [their] activity . . . not the motivations behind the effects'. Physical geographers see their research as scientific in the double sense that (i) it is about a really existing non-human world whose operations are absolutely or relatively autonomous from society, and (ii) that it actually or potentially represents that world as it really is. As Bruce Rhoads (1999: 765) puts it in relation to (i), quoting the philosopher Ian Hacking (1996: 44), 'At the most fundamental level . . . physical geographers . . . subscribe to the general . . . sentiment that "there is one world susceptible of scientific investigation, one reality amenable to scientific description, [and] one totality of truths equally open to scientific inquirers . . ." '. In relation to (ii) we can observe that some version of the so-called 'correspondence theory of truth' is involved, wherein scientific knowledge is seen to 'mirror' the biophysical world. We might also note that physical geographers are realists in the two senses I identified above about all three of the principal meanings of the term nature (as laid out in Chapter 1). Not only do they believe in the reality of the non-human world, they also believe it has an essential character amenable to discovery. And they are often interested in the inherent forces – like energy fluxes – that structure and connect different elements of the biophysical environment.

On one reading, this suggests that a happy division of labour exists within geography vis-à-vis nature. According to this view, physical geographers investigate the 'true nature' of the non-human world, while human geographers examine the socially variable representations of, and actions upon, those things we call natural (human and non-human). Meanwhile, environmental geographers do a bit of both, depending. On this interpretation, the discipline of geography offers us a truly comprehensive understanding of nature, ranging from nature in itself to the discursive and material constructions that societies impose upon it.[5] This positive interpretation contrasts with the one I put forward above, of a discipline whose knowledges of nature are divided. But this sanguine viewpoint on how geographers carve up the study of nature is too simple. The scientific and realist credentials claimed by physical geographers should not, I contend, be accepted at face value. In keeping with arguments presented in Chapter 1, I suggest that they be seen as moves in a high-stakes game. In this game, many actors and institutions are vying to have their knowledge of those things we call natural accepted (and acted upon) by significant sections of society. Though it may well be the case that physical geographers *do* produce reliable knowledge about the non-human world relatively free from bias, my interest in this chapter is in how and why they *claim* to do so.

In this light, it is useful to speculate about why contemporary physical geographers might wish their research to be seen as both scientific and realist. The following Activity question is designed to get you thinking about what is at stake when physical geographers matter-of-factly tell students, other researchers and non-academic groups that the environmental knowledge they produce is (or aspires to be) more or less truthful.

ACTIVITY 4.1

Imagine that you are a physical geographer employed full-time by a university. Your main specialism is landslides. Because it's expensive to undertake, your research requires funding from outside bodies (like government agencies). It also has an applied element because many human populations live in potential landslide areas. Why would it be important to you to emphasise, or at least not downplay, the scientific and realist character of your research?

One obvious answer to the activity question is as follows: claiming that one can produce an accurate (i.e. scientific) account of a real environmental phenomena (in this case landslides) is a way of gaining trust – the trust of funding bodies and policy-makers, for example. If one were to deny that landslides really existed, or if one were to be perceived as an 'unscientific' researcher, then it's unlikely that one's research would be funded, let alone believed. Since most people in academia and the wider world are ultimately realists, then the claim to be investigating reality 'scientifically' becomes the chief means by which researchers can establish a privileged status for their knowledge. As Gieryn (1983) pointed out, science is a normative term that allows those who appropriate it to perform 'boundary work'. To say that one is a scientist is to distinguish oneself sharply from those people who produce supposedly 'lesser' knowledge (i.e. non-scientists). In Demeritt's (1996: 485) words: 'Debates about science . . . are debates about what will count for real knowledge and whose voices will be heard in struggles to define it'. Or, as Derek Gregory (1994: 79) put it in a similar vein, 'Science is a weasel-word . . . [I]t is much used (and abused) as a term of approbation or condemnation, made to stand for a system of knowledge to which we are enjoined to aspire'.

I am not suggesting that physical geographers are party to some grand conspiracy – one that involves using the label science for purely self-serving reasons! This suggestion would be cynical and unjustified. I am simply asking why a commitment to the ideas of science and realism are so deeply insinuated into their self-understanding. The issues of trust and boundary-work aside, there are other reasons why physical geographers might wish to perpetuate the view that their research is scientific and realist (these can be gleaned from a scan of recent literature). In the first place, we live in an era of immense public and governmental concern about human alterations of the non-human world. This provides an ideal opportunity for physical geographers to meet a growing demand for accurate understandings of anthropogenic environmental change. If such understanding cannot be arrived at, we arguably risk formulating incorrect environmental policies or we might fail to identify biophysical problems early enough (see Graf 1992). Second, human alterations and applied research aside, there are many aspects of the physical environment that we still do not have a good understanding of in their own right, such is their complexity – like ocean–atmosphere couplings. Third, it's arguable that environmental research that is too analytical – which intellectually severs the connections between interrelated environmental phenomena – is 'unrealistic'. In this context, physical geography's synthesising ambitions appear necessary for a proper understanding of how the non-human world works – which is why Slaymaker and Spencer (among others) lament the subdisciplinary separations of geomorphology, hydrology, biogeography and climatology.

In sum, we can adduce many good reasons for believing that accurate knowledge of the biophysical world is a desirable (and achievable) thing. Governments and the wider public are willing audiences for researchers who can offer their 'expert' insights into the 'real nature' of natural and humanly altered environments. All this is reinforced when we look at the case against epistemological anti-realism (or what's sometimes called 'relativism' or 'conventionalism'). In relation to disciplines that classify themselves as sciences, this case has been prosecuted most vigorously by historians and philosophers of science. Expressed in simple terms, relativists argue that all knowledge about nature (including scientific knowledge) is contingent and constructed, rather than a true reflection of the reality it apparently represents. What counts as a truth about what we call nature is thus seen as relative to the perspective of the viewer or investigator – including even scientists. I will discuss one variant of relativism in the next section, but for now we can identify some apparently strong arguments against it. Physical geographers have rarely felt obliged to make these arguments (for reasons to be explained), but they clearly bolster the idea that their research is realist. First, one can argue that relativists cannot be correct because biophysical reality will ultimately contradict any false representations of its true character. Second, even if one concedes that scientific knowledge is always at some level a reflection of scientists' mindsets (see the next section), it remains the case that that this knowledge is not purely self-referential. Instead, it is always *about something other than itself*: namely, a world that exists separately from it. If this were not the case then researchers would have nothing to research! And since that world cannot be easily altered, let alone constructed (e.g. most people would agree that one cannot 'construct' the river Nile, only representations of it), it follows for realists that their knowledge is always more than a groundless fabrication. This knowledge has real referents that condition and constrain how those referents are represented by researchers (for a lucid introduction to the relativism–realism debate in science see Kirk 1999 and Okasha 2002: ch. 4).

THE SOCIAL CONSTRUCTION OF SCIENTIFIC KNOWLEDGE

Before I discuss the investigative procedures that, broadly speaking, give physical geographers confidence in the realism (in sense (ii) above) of the knowledge they produce, I want to explore the possibility that this confidence is misplaced. I mentioned above that physical geographers have rarely felt compelled to defend their methods of environmental investigation in any formal sense. For the most part, they take it as given that physical geography *is* a science, leaving room only for a debate over what *kind of science* it happens to be (as we'll see in the next two sections). This is odd for two reasons. First, over the past two decades, a field of study called the sociology of scientific knowledge (or SSK) – sometimes known more generally as science and technology studies (STS) – has questioned the objectivity of scientists' findings. Examining the activities of a whole range of different scientists operating in different disciplines and institutions,[6] SSK researchers have suggested that even 'scientific knowledge is made up, just like fairy tales and nursery rhymes' (Demeritt 1996: 484). SSK has been central to the so-called 'science wars' to which I referred in the Preface and, in the eyes of its detractors (e.g. Gross and Levitt 1994), is 'anti-science'. Second, the image of scientific knowledge has been tainted in recent years by a series of public-health scares – like avian flu and BSE. As I observed in Chapter 1, scientists' unawareness of, or uncertainty about, these manufactured hazards have shaken public faith in their expertise. Given this twin context, one might have expected physical geographers to launch a defence of the scientificity of their research. Yet, as I've said, the reality is that these geographers rarely bother to debate the issue of science at all. Why is this? The reasons, I think, are purely disciplinary. Historically, physical geographers – as Richard Chorley famously quipped – have instinctively reached for their soil augers when debates threaten to get too philosophical and abstract. In addition to this, the de-naturalising turn of contemporary human geography could be ignored by physical geographers so long as there was no suggestion that their own representations of nature were constructions. In other words, because of their immersion within a distinct disciplinary context, physical geographers have rarely been obliged to spell out how and why the knowledge they produce is a faithful depiction of biophysical realities.

In recent years this has changed. In a series of important essays, David Demeritt – a human-cum-environmental geographer at King's College, London – has applied the insights of SSK to physical geographers' research (Demeritt 1996; 1998; 2001c; 2001d). In other words, Demeritt has extended the 'nature-sceptical' sensibilities of many human geographers to that part of geography whose reputation rests on its claim to tell us how nature 'really works'. He has been able to do this because, unlike most human geographers, he has an earth-science background (including expertise in climate modelling developed when he worked for Environment Canada). What makes Demeritt's research compelling is that he scrutinises the *practice* of physical geographers. In other words, his arguments about the construction of scientific knowledge are not made at the philosophical level but are demonstrated empirically. For this reason, his arguments are difficult for physical geographers to ignore – as indicated by Schneider's (2001) terse response to one of Demeritt's papers (on how scientific knowledge about global warming is constructed). I mention Demeritt's physical-science expertise and his focus on scientific practice because some earlier criticisms of physical geographers scientific approach *were* ignored. For instance, in 1993 Gillian Rose drew upon the work of feminist historians of science to argue that physical geographers' knowledge is inherently masculinist. But because her arguments were theoretical they arguably lacked the precision required to persuade physical geographers that she might be on to something important.

I cannot do justice to the richness of Demeritt's arguments here. So let me open just one window onto them, first by summarising the main theses of SSK, and then by showing how Demeritt applies SSK to one aspect of environmental research. SSK originated with the pioneering studies of

David Bloor, Harry Collins, Barry Barnes, Bruno Latour and Steve Woolgar in the 1970s – five philosophers, sociologists and historians interested in how scientific knowledge is generated and legitimated. SSK researchers hold to the 'symmetry principle': that is, the idea that scientific beliefs held to be true should be analysed in the very same, socially constructionist terms as those held to be false (see Figure 4.1). These researchers argue that if we're to understand the truths about nature discovered by scientists we need to look at the scientific community itself not the natural world. In other words, SSK researchers maintain that scientific facts do not 'speak for themselves' but are spoken for and stage-managed by scientists. This is not to suggest that scientists consciously try to deceive people or to wilfully concoct erroneous findings. Rather, SSK researchers argue that it is the unconscious, tacit and taken-for-granted elements of scientific practice – like the routine ways in which data is gathered and analysed – that inevitably produce constructed rather than realistic knowledge. Indeed, Collins (1985) maintains that scientists can never really know whether their understanding of the world is accurate or not. For instance, where disagreements between scientists arise it is unclear whether the methods used to investigate reality were flawed or whether the data gathered is actually correct and contradicts prevailing (and erroneous) scientific beliefs. (Good introductions to SSK and STS have been written by Hess [1997] and Sismondo [2003]).

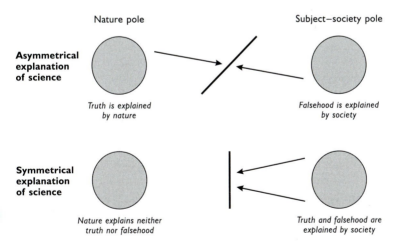

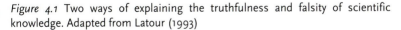

Figure 4.1 Two ways of explaining the truthfulness and falsity of scientific knowledge. Adapted from Latour (1993)

Drawing upon SSK arguments, Demeritt, in one of his published essays, engages in an auto-critique. He reflects upon his own 'scientific research' into the climatic impact of stratospheric volcanic aerosols in north-eastern USA – research conducted in the early 1990s prior to his engagement with the SSK literature. In this research, Demeritt sought to identify possible causal links between volcanic eruptions during the nineteenth and twentieth centuries and climatic variance in the New England states. As part of this he used long-run temperature data transcribed at various weather stations and searched for any volcanic signal with the 'noise'. Reasonable as this correlation exercise may seem, it presumes that the temperature data is a reliable account of real temperature changes over time. But, as Demeritt notes, this cannot be proven: the contingencies of where, when, and how carefully temperature readings were taken in previous decades is unknown. Demeritt simply had to *assume* that the temperature data were reliable. In effect, the data was a 'black box': its veracity could never be demonstrated. And even if Demeritt knew how rigorously taken temperature readings were up to two centuries ago, how, he asks, would we know what level of rigour is acceptable? Are ten temperature readings per day per weather station enough? How many weather stations are needed to give a true depiction of the climatic conditions of the north-eastern seaboard? The answers to these questions, Demeritt, argues, are not dictated to us by the natural environment. They are a matter of judgement. And herein lies the problem: if the temperature data that Demeritt used are the only direct evidence we have of 'real temperature' then we can never know whether the data reflects reality or reality contradicts the data. Since we cannot go back in time and check, we left in a position of having to take this data at face value as if it corresponds to past climate.

Demeritt's preoccupation with the quality and quantity of data used in his research speaks to a wider issue in physical geography (and indeed all fields of research). Physical geographers routinely use data-sets created by other researchers and organisations and take it on trust that these data sets possess a fair degree of truth-value. Equally, these geographers generate their own original data. In light of Demeritt's analysis, one could raise questions about how these geographers know when an 'acceptable' quantity and quality of data has been gathered about any given environmental phenomena. More generally, the insights of SSK would lead us to examine each and every stage of a physical geographer's research – from the theories and hypotheses used to the equipment utilised to the way data is analysed.

It's important to stress that Demeritt does not see his constructionist approach to scientific knowledge as anti-science. Rather, his aim is to get physical geographers and other earth scientists involved in a more honest discussion about the status of the knowledge they produce. Whether physical geographers will formally engage with the claims of SSK remains to be seen. But Demeritt's research has opened the door for this possibility. What's more, as we'll see towards the end of the section 'Understanding biophysical reality' on pp. 202–18, a small number of physical geographers are already thinking along similar lines as Demeritt, albeit without reference to SSK.

PRODUCING REALISTIC ENVIRONMENTAL KNOWLEDGE

In the absence of much fundamental self-questioning about the actual or potential accuracy of the environmental knowledge they produce, most physical geographers have focused on questions of *method* in order to flesh out what is scientific about their research. As fluvial geomorphologist Keith Richards argues, 'science as an activity or entity seems to be defined less by what it is, than by how it is done' (Richards 2003a: 25). Virtually all physical geographers accept that the non-human world is ontologically real and, for the most part, ontologically different in character from the human world. Likewise, virtually all physical geographers accept the possibility that the non-human world is knowable in relatively unbiased ways. This is why such discussions of science as there have been in physical geography are usually discussions of method. As Schumm (1991: 2) observes, 'in the minds of most scientists, it is the method employed in carrying out their research that distinguishes science from other human endeavours'.

Scientific method in physical geography

I use the term 'method' in the widest sense to mean a set of steps followed in the investigation of the biophysical world – that is, 'method as a way of doing anything according to a regular plan' (Haines-Young and Petch 1986: 10). In other words, my concern here is not with specific quantitative and qualitative methods of data-gathering or analysis (like soil-corers or carbon-dating). What Robert Merton (1942) famously called the 'universality of science' derives, in his view, from the fact that any suitably trained individual can follow these steps – regardless of gender, colour, class etc. So what are these steps?

ACTIVITY 4.2

Following on from the previous Activity, imagine that you wish to identify the cause/s of landslides in an area of the world where landslides have been little researched. What, in your view, would be the principal steps you would take in order to undertake the analysis?

Your answer will, hopefully, have included some or all of the following. The starting point is an environmental phenomena as yet unexplained. As Bird (1989: 2) noted, 'scientific method starts with some kind of problem: we might go as far as to say that problem orientation is the *raison d'être* of scientific inquiry'. This problem-focus typically takes the form of posing a 'why?' or 'how?' question which needs to be answered – in this case why landslides occur(red).[7] Second, some initial, preliminary observations of the landslides in question will produce some ideas that might explain why they occurred. These ideas do not, of course, emerge purely from observation. Rather, preliminary observations are already structured by the fact that the researcher knows the research literature on landslides. S/he is thus already familiar with the principal explanations of landslides based on studies undertaken elsewhere in the world. They are likely to draw upon this knowledge when speculating about what caused the landslides under investigation. As a result, the researcher will specify a possible explanation in the form of a model or theory (and, in many parts of physical geography, a law). In basic terms, a model is a simplified representation of reality that aims to depict the key causal variables or interrelationships at work (or the 'signals in the noise'). A theory is usually a more sophisticated and detailed attempt to offer a rational explanation of reality and comprises a set of consistent, logical statements that would account for the existence of the phenomena under investigation. A law, meanwhile, describes a consistent relationship (deterministic or probabilistic) between two or more variables that is more or less universal in nature.[8] Equipped with a model, theory or law, the researcher might then derive some empirically testable hypotheses.[9]

In this case, the hypotheses would be statements about such likely causal factors as slope gradient, soil moisture content, soil porosity and the like, as well as the relationships between them. These hypotheses would then be subject to empirical scrutiny through the classification, measurement and further observation of the landslides (and perhaps of landslides yet to happen i.e. soil and vegetation on slopes likely to suffer failure in the future). This might involve conducting controlled field experiments, running a computer simulation, or even undertaking some laboratory experiments (e.g. saturating a scale model of the terrain with water and observing the results). In turn, the data would be analysed in light of the hypotheses proposed. Such analysis may lead to the confirmation, amendment or even refutation of some or all of the hypotheses put forward. If necessary, further testing of new or altered hypotheses may be undertaken leading, hopefully, to a robust theory or model that is applicable to the landslides in question (Figure 4.2).

This is, of course, an ideal-typical account. In practice, most physical geography research does not follow this neat and tidy step-by-step procedure. Twidale's (1983: 55) confession doubtless still holds true: 'the so-called methods are in reality haphazard, intuitive or even serendipitous . . . when scientists have attempted to record the sequence of events leading to discovery, they describe what they think ought to have been done rather than what was indeed the case'. This is, perhaps, why Schumm (1991) prefers to talk of a fairly loose 'scientific approach' rather than a strict method (while Rhoads 1999 identifies no less than seven 'ways of knowing' in physical geography). Even so, the investigative procedure sketched in the previous paragraph loosely *approximates* the conduct of physical geographers' research in very many cases. The way I've described this procedure conceals a number of methodological issues and principles that have been given separate labels but which, in reality, are mixed and matched in any given physical geographer's research practices. These issues and principles warrant a brief mention because they all relate to how scientific method can, in the eyes of physical geographers, yield more realistic and accurate knowledge of the biophysical world. Again, I'm forced to simplify and generalise in order to tease out some core dimensions of how many physical geographers undertake research.

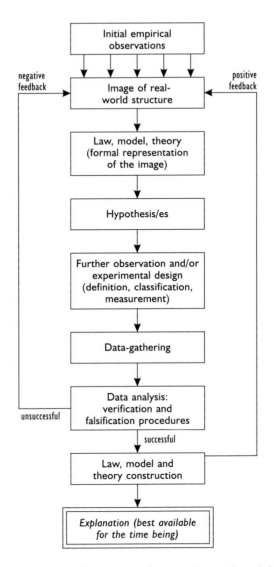

Figure 4.2 Investigating reality: a scientific procedure. Adapted from Harvey (1969)

Issues and principles of scientific method

There are six issues and principles of physical geographer's method that I want briefly to consider. First, induction is part and parcel of physical geographers' investigative practice. But it is accepted that it cannot be its sole basis. Literally defined, induction means (i) forming an impression about the real world from a process of pure, presuppositionless observation and (ii) making generalisations about a class of phenomena on the basis of specific observed set of the same phenomena. Induction is thus the idea that generalisations can be made (inferences) if a specific set of facts are allowed to 'speak for themselves' through objective observation. Though physical geographers do undertake preliminary observations of the biophysical world at the start of any analysis, it is understood that this can never be a 'pure' process leading to equally pure hunches about what one has observed and what explains it. Rather, as Popper pointed out decades ago, at the start of any research project scientists have (i) already decided what kinds of phenomena are worth observing and (ii) they already have an idea why these phenomena are as they are on the basis of previous research. The physicist Werner Heisenberg (1958: 12) expressed it thus: 'What we observe is not nature itself, but nature exposed to our method of questioning'. In this sense, any initial observations of the physical environment are seen to be theory-laden (though not necessarily theory-determined). What's more, generalising from a limited set of observations is seen as always precarious because future observations might undermine the generalisations made.

Second, there is frequently a deductive dimension to scientific method in physical geography. Burt (2003a: 59), for example, claims that 'these days, physical geography is firmly ensconced as a deductive science'. Deduction involves reasoning from known laws, theories or models to as yet unknown or unresearched phenomena likely to be explicable in terms of those laws, theories and models. The researcher deduces what did, should or will occur on the basis of established 'empirical truths' (yielded by previous research), established 'logical truths' (e.g. those specified by mathematicians and statisticians) and factual information about the case being researched ('initial conditions': see Box 4.2).[10] Even so, Marshall (1985) is right to point out that all empirical research (in physical geography and beyond) is ultimately inductive in the obvious sense that it rests on data that may be contradicted by future studies. Third, when through a combination of induction and deduction a physical geographer has formulated a plausible idea that might account for what s/he's observed in the landscape, it is, these days, accepted that *multiple working hypotheses* are preferable to a single ruling hypothesis (Chamberlin 1965). The reason for this is that testing multiple hypotheses maximises the chance of identifying the correct ideas about what is being studied, while also speeding up scientific discovery (see Figures 4.3 and 4.4). The paper by Battarbee et al. (1985) is a classic example of multiple working hypotheses being utilised in a research project.

Box 4.2 THE DEDUCTIVE-NOMOLOGICAL MODE OF SCIENTIFIC EXPLANATION

The deductive-nomological (or 'covering-law') approach to explaining the world was first codified by the so-called Vienna School of the 1920s and 1930s. This group of philosophers and mathe-maticians argued that science can be demarcated from non-science because it only deals with two kinds of truths: namely, empirical truths (i.e. those established by unbiased observation) and logical truths (like 1+1=2). Karl Popper subsequently argued that the former are best arrived at through a process of falsification not verification. Together, these two kinds of truth are expressed as scientific models, theories or laws. For the Vienna School these ought to be absolutely or relatively universal, covering phenomena as yet unobserved in so far as those phenomena are the same as those upon whose observation existing models, theories and laws are based. This means that a practising scientist (like a landslide researcher) can use these models, theories and laws in new empirical settings rather than having to create new ones each time they undertake research. Put differently, this presumption of the relative or absolute universality of scientific knowledge 'enable[s scientists] . . . to connect together [their] . . . knowledge of separately known events, and to make reliable predictions of events as yet unknown' (Braithwaite 1953: 1). As two physical geographers express it in relation to their research area: 'It is of limited interest to know why the sediment load of the River Rhine varies as it does; but it is a different matter if knowledge of the Rhine's sediment-carry characteristics help us to understand the unifying principles controlling the sediment loads of the River Exe, the River Rhine and the Amazon river' (Favis-Mortlock and de Boer, 2004: 164). A deductive-nomological explanation takes the following form:

$L_1, L_2 \ldots L_n$	(Laws, theories and models)
$T_1, T_2 \ldots T_n$	
$M_1, M_2 \ldots M_n$	
+	
$\underline{C_1, C_2 \ldots C_n}$	(Initial conditions)
E	(Past, present or future event/s)

Here, a set of empirical events can *necessarily* be described, explained and/or predicted from a set of well-confirmed laws, theories or models coupled with factual information about the local conditions prevailing at the site where the explanation or prediction applies. For instance, if a hydrologist has a set of general laws about soil porosity and water throughflow, plus information about the local soil type and its antecedent moisture content, they might be able to both explain *and* predict why and whether overland flow occurs during a particular rainstorm as opposed to subsurface flow. In physical geography, it is frequently the case that deductive-nomological explanation takes a probabilistic rather than strictly deterministic form because of the open-systems nature of the biophysical world. What's more, in practice deductive reasoning is bound up with deductive and abductive reasoning (see Box 4.5 for a discussion of the latter).

The deductive-nomological form of explanation is usually equated with a *positivist* view of science. Physical geography is often considered to be positivist (and human geography was once positivist according to some). But in my view this label has become meaningless through overuse. The term lacks clarity of meaning because it has been used indiscriminately in the literature over the years.

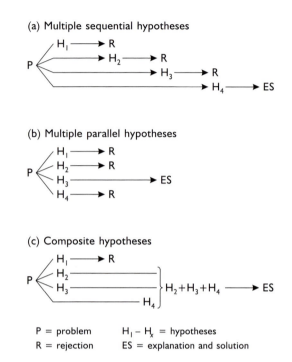

Figure 4.3 Hypotheses in physical geography. Adapted from Schumm (1991)

Fourth, because of Popper's well-known 'critical rationalist' approach, many physical geographers accept that *verification* is a logically insufficient basis for the testing of hypotheses. Verification involves identifying evidence that confirms a particular hypothesis. Popper argued that this was a logically flawed approach to testing because while 10,000 observations might confirm any given hypothesis, the 10,001st might falsify it. Popper thus favoured falsification as a testing procedure, wherein the researcher actively looks for evidence that *disproves* a hypothesis. In physical geography, Haines-Young and Petch (1986) and Richards (2003a) strongly advocated critical rationalism on the grounds that it is an efficient and rigorous way of identifying true and false hypotheses (and thus theories, models and laws). Fifth, most physical geographers are cautious about the truth-value of the knowledge they produce. Human geographers often hold to the unthinking stereotype that their physical counterparts are unsophisticated epistemological realists who believe their knowledge is True with a capital T. But

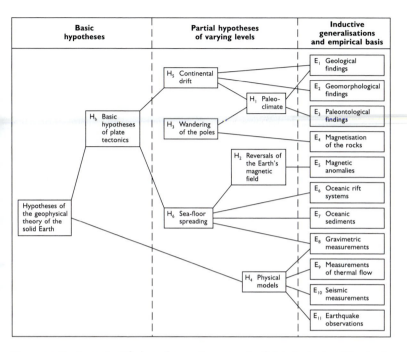

Figure 4.4 Testing scientific hypotheses: the example of plate tectonics. Here basic hypotheses are translated into empirically testable ones which, in turn, are assessed against various bodies of evidence. Reproduced from von Engelhardt, Zimmerman and Fischer (1988)

this is grossly unfair. In practice, most physical geographers see the knowledge they produce as *provisionally true* not absolutely so: as the best available representation of how the biophysical world operates. This is concordant with the well-known idea that science is 'organised scepticism': a consistent procedure for testing, amending and improving existing knowledge about the material world. Finally, it is wrong to assume that *all* physical geographers adhere to a 'mimetic' or 'correspondence' view of knowledge. This view, as I noted earlier, assumes that scientific knowledge reflects reality as in a mirror. It thus sidesteps the idea that knowledge is a filter that strains and sieves out sense-data such that we can never know reality 'as it really is' (an idea I discussed at length in Chapter 1). But some physical geographers operate with *coherence* and *utilitarian* (or *instrumentalist*) views of the knowledge their research produces. The former is the view that knowledge of reality is likely to be correct if it is consistent with existing bodies of knowledge, as well as appearing to withstand empirical testing. The latter is the view that if knowledge of the physical environment is practically useful – as when a theory or model of landslides successfully predicts when future landslides will occur – then this is the main criteria of its worth.

The above discussion in no way does justice to the raft of methodological issues and protocols that have 'seeped' into the research culture of physical geography (excellent discussions are offered by Haines-Young and Petch 1986, Inkpen 2004 [chs 2–5] and Richards 2003a). But my main point, I think, holds true: namely, that most physical geographers question not *whether* the physical environment is knowable but how best to grasp its real nature. As Raper and Livingstone (2001: 237) put it, 'physical geographers . . . see representation as connecting real entities and mental concepts within a framework of realism about the external world'. Symptomatic of this is the fact that the ideas of Paul Feyerabend (1924–94) have cut little ice with physical geographers. Feyerabend, an outspoken historian of science, sought to expose the 'myth of method' in the natural sciences. In his provocative book *Against Method* (1975) he argued that the protocols of scientific method have been consistently flouted by practising scientists. For Feyerabend scientists are methodological pluralists who adhere to no one investigative procedure, however much they may claim to do. As Haines-Young and Petch (1986: 99) summarise: 'the only methodological rule that can be defended in all circumstances [for Feyerabend] is . . . that *anything goes*' in science. Feyerabend, in short, saw the method-myth as a ruse used to persuade society that scientific knowledge can offer uniquely objective insights into how the world works. Rather as physical geographers have ignored the challenge of SSK thus far, so too have they bracketed Feyerabend's objections for the most part (see Haines-Young and Petch 1986: ch. 6).[11] It's not difficult to understand why: after all, Feyerabend throws into question the idea that science is about fact not fiction, truth not lies. This said, many (if not all) physical geographers would accept that their investigations involve more than the exercise of a cool, dispassionate rationality that obeys only the evidence gathered. See Box 4.3 for a fascinating example.

Box 4.3 INTELLECTUAL DISPUTES IN PHYSICAL GEOGRAPHY

Physical geography, like any other field of intellectual inquiry, is characterised by disputes between practitioners over any and all aspects of the process and the results of research. Some of the most intense disputes relate to the quality and significance of evidence: what does a given body of evidence tell us about the biophysical world? If physical geographers were robots rather than what they are (i.e. thinking, feeling people) then one might imagine disputes over evidence could be quickly and 'rationally' resolved. But the reality is more complicated. Sugden (1996) provides a fascinating example. In the mid-1990s there were two rival accounts of the history of the East Antarctic Ice Sheet involving physical geographers and geologists. One school of thought (the 'dynamic' school) insisted that the ice sheet largely disappeared during the late Pliocene period. Another school (the 'static' school) argued that the ice sheet had been remarkably stable, even during periods of naturally increased temperature. Both schools presented evidence to support their respective cases, and it was clear that deciding which one was correct was of more than academic importance. After all, if we are currently experiencing 'global warming' then it's important to know if the 'dynamists' are right since large sea-level rises (among other things) would be a likely future scenario worldwide such is the amount of H_2O locked-up in Antarctica. Sugden examines how those in the dynamist camp – which was dominant through the 1980s – 'dealt with' the evidence presented by the stabilists. That evidence was not only wide ranging. It also challenged the main factual basis of the dynamist perspective, the so-called Sirius Group deposit evidence taken from thirty-three high-altitude sites in the Transantarctic Mountains. Diatom analysis of the Sirius tills and gravels suggested that temperate forest had existed during the late Pliocene, akin to that in Patagonia today. Against this, the stabilists presented new evidence in 1994 that suggested that the diatoms were not indigenous to Antarctica and were airborne 'imports' to Antarctica. Confronted with this troubling new evidence, Sudgen shows how the dynamists used several strategies to 'protect' their perspective. One of these was to cast doubt on the credibility of the stabilists' evidence. For instance, one piece of evidence related to volcanic ash which had, apparently, remained relatively undisturbed through the Pliocene, casting doubt on the dynamists' ideas of major environmental change in the region. In response, Sudgen (1996: 499) reports two dynamists casting around for reasons to dispute the apparently indisputable, suggesting that the ash deposits may overlay till that *had* been moved during a melting phase. Overall, Sudgen shows that intellectual disputes in physical geography are not cleanly resolved by recourse to 'the facts'. Instead, often entrenched perspectives prove difficult to alter because researchers have invested time, money and their reputations in developing them.

UNDERSTANDING BIOPHYSICAL REALITY: SOME KEY DEBATES

While most physical geographers express faith in the rigour of their investigative procedures, this is not to suggest that they share the same epistemological and ontological beliefs. It is one thing to assume that there is a 'biophysical reality independent of the human mind' (Phillips 1999: 7). But it's quite another to agree on how, broadly speaking, we can come to know that world and how, broadly speaking, that world is structured. All researchers – physical geographers or otherwise – have ontological and epistemological beliefs (as indeed do all people: see Box 4.4). These beliefs form the context for the investigative steps discussed in the previous section. They are, if you like, a researcher's 'bedrock assumptions'. Ontological beliefs are general beliefs about what is real (or what exists). Epistemological beliefs are general beliefs about how we, as humans, can come to know reality. Though I've not used the term thus far, I've already explained that most physical geographers are *materialists* in the ontological sense. That is, they believe in the existence (or reality, hence the term 'real') of a physical world independent of, or at least irreducible to, any given set of human perceptions of or actions upon that world. But not all materialist ontologies are the same, as we'll see below.

Box 4.4 ONTOLOGIES AND EPISTEMOLOGIES

Whether they know it or not, everyone has ontological and epistemological beliefs. Though ordinary people rarely reflect upon these beliefs, professional researchers tend to do so periodically, if not frequently. It is useful for researchers to be explicit about their ontological and epistemological beliefs so that they can be scrutinised and perhaps even challenged. Ontological beliefs specify what is real (or what exists), while epistemological beliefs specify how we can know reality. Broadly speaking, people who believe that there is a real world independent of human perception and cognition are ontological materialists (or ontological realists). Conversely, those who – like some discursive constructionists (see Chapter 3) – believe that human ideas determine what is real for us are ontological *idealists*. Likewise, we can draw a distinction between ontological *atomists* and ontological *holists*. The former believe that reality is comprised of discrete parts that interact, while holists maintain that the operation of parts depends upon their relationships with all others within an integrated system. In practice, there are many variants of materialism, idealism, atomism and holism. For instance, while some materialists believe that the non-human world is inherently orderly in its behaviour, others believe it is unstable and chaotic. Epistemologically, people who believe that 'seeing is believing' are empiricists. By contrast, those who believe that much of reality is invisible to the eye (like gravity or the social norms that structure how men and women interact) are non-empiricists. Ontological and epistemological beliefs underpin all research. For instance, if one is an ontological holist then this will profoundly affect how one classifies observed phenomena in any investigation. Since one cannot readily 'cut the biophysical world at the joints' (as an atomist would suppose one could), then the epistemological act of deciding what conceptual boxes to use becomes important since these boxes may falsely separate what, ontologically, are 'internally related' phenomena.

Epistemologically, physical geographers are a diverse group. For many human geographers, the stereotypical Wellington-boot-wearing physical geographer is an *empiricist*. That is, s/he believes that we can only truly know what we can *see* with our eyes (either directly or through the use of photographs, microscopes, recording devices etc.). But this stereotype is wide of the mark, as we'll also discover momentarily.

Why, it might be asked, am I straying into philosophical waters at this point in the chapter? Aren't I moving away from discussing the 'real environments' that physical geographers are interested in? My answer to the latter question is 'no' and to the former is twofold. First, the epistemological and ontological debates among physical geographers reveal some of the key differences in how they understand the 'real nature' of the nonhuman world. These differences are not a denial that there is a real physical world 'out there' but, rather, broad disagreements about how we might know that world and how it is structured. Second, these debates link directly to the methodological issues discussed in the previous section 'Producing realistic environmental knowledge'. In other words, these debates show us that for physical geographers a 'proper' understanding of the physical environment is not just a matter of method but also of the wider assumptions guiding the practical use of method. What follows is by no means a complete inventory, but it gives a sense of the sophisticated debates about the biophysical environment current in physical geography among a strong minority of researchers. These researchers are especially evident in physical geography's largest subfield, geomorphology.

Ontological issues

There are four ontological debates worth mentioning that cut to the heart of how physical geographers understand the 'nature' of the biophysical world. Though these debates are about the most appropriate (i.e. realistic) way to think of that world, I suggest that – in keeping with this book's overall argument – that we see them as rival *imaginaries* of the biophysical world. I should preface my presentation of these debates by pointing to one corollary of the fact that physical geography is, in large measure, a field science not a laboratory science. Physical geography must deal with a subject matter that is *multi-scalar* – stretching from the smallest spatio-temporal scales to the largest (see Figure 4.5). Although, physical geographers narrowed their focus after the 'spatial science' revolution of the 1950s and

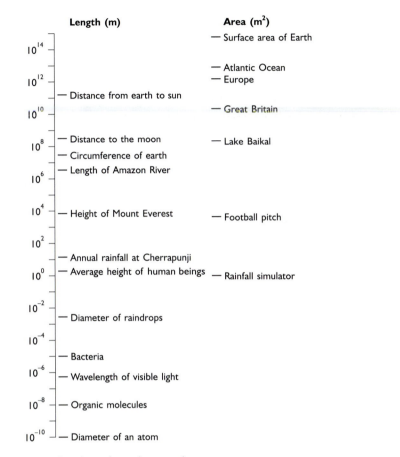

Figure 4.5 Spatial scale in physical geography

1960s, recently there has been a return to the study of larger-scale phenomena – like El Niño and global warming (see Spedding 2003). The space in between the two (meso-scale studies or what's sometimes called 'regional physical geography') is still only lightly populated. Thus physical geography is currently strung out between in-depth, micro-scale studies of environmental processes and their effects, and macro-scale studies of environmental change.

First, there is a debate over where the biophysical world is stratified. In this context, the term stratification refers to an ontological 'layering' of the non-human world. If that world is layered then this means that what holds true at one spatio-temporal scale may not hold true at a higher or lower one. Here each strata of biophysical reality is composed of, but not reducible to, entities that exist at smaller scales. In a classic paper, Schumm and Lichty (1965) implied that stratification was part and parcel of the world that physical geographers study. They argued that the dependent and independent explanatory variables changed depending on the spatio-temporal scale in question (see Table 4.1). Four decades on, physical geographers remain uncertain how far analyses conducted at one scale are applicable to other scales. Many engaged in small-scale research into environmental processes have tried to upscale their findings. But, as Stephan Harrison (2001) argues, this is implicitly *reductionist* ontologically speaking. That is, it suggests that 'the real essence of an object of inquiry can be seen at the microscopic "fundamental" scale' (2001: 330). In the words of another commentator, 'Reductionist studies . . . have become the *modus operandi* of much scientific research [and] often take place at very detailed spatial and temporal scales. Reductionism functions by studying small systems in detail in order to aggregate the information . . . about a broader system' (Barrett 1999: 709). Against this, it is possible to argue that larger-scale environmental phenomena have so-called 'emergent properties' that cannot be 'read off' from the properties that exist at smaller scales. For instance, if one wants to know why mountain ranges form over long time-periods is it necessary to understand the molecular properties of all the rock types in those ranges? Some would say not. The debate over stratification and reductionism remains ongoing and unresolved. It is, as Sudgen et al. (1997: 193) aver, 'a nut that must be cracked' (see Burt 2003b).

Second, there has been a related debate over whether physical geographers study so-called 'natural kinds' or not. A natural kind is any element of the real world that is possessed of the following two qualities. First, it is ontologically different and distinct from other elements (even though it may be related to those other elements in practice). Second, it retains its physical integrity regardless of the specific circumstances in which it exists. Thus, a piece of granite might be considered a natural kind if it can be shown that it is unlike (qualitatively different from) other kinds of rock and that it remains granite whether it is found on a scree slope or at the bottom of the ocean. It is often said that the 'hard sciences' (like physics) study natural kinds: that is, the basic 'building blocks' of the biophysical world. But does a field science like physical geography study natural kinds too? The Activity on page 209 invites you to construct an answer to this challenging question.

Table 4.1 (a) The status of river variables during time spans of decreasing duration. Reproduced from Schumm and Lichty (1965)

River variables	Status of variables during designated time spans		
	Geological	Modern	Present
1 Time	Independent	Not relevant	Not relevant
2 Geology	Independent	Independent	Independent
3 Climate	Independent	Independent	Independent
4 Vegetation (type and density)	Dependent	Independent	Independent
5 Relief	Dependent	Independent	Independent
6 Palaeohydrology (long-term discharge of water and sediment)	Dependent	Independent	Independent
7 Valley dimensions (width, depth, slope)	Dependent	Independent	Independent
8 Mean discharge of water and sediment	Indeterminate	Independent	Independent
9 Channel morphology (width, depth, slope, shape, pattern)	Indeterminate	Dependent	Independent
10 Observed discharge of water and sediment	Indeterminate	Indeterminate	Dependent
11 Observed flow characteristics (depth, velocity, turbulence, etc.)	Indeterminate	Indeterminate	Dependent

continued

Table 4.1 (b) The status of drainage basin variables during time spans of decreasing duration

Drainage basin variables	Status of variables during designated time spans		
	Cyclic	Graded	Steady
1 Time	Independent	Not relevant	Not relevant
2 Initial relief	Independent	Not relevant	Not relevant
3 Geology	Independent	Independent	Independent
4 Climate	Independent	Independent	Independent
5 Vegetation (type and density)	Dependent	Independent	Independent
6 Relief or volume or system above base level	Dependent	Independent	Independent
7 Hydrology (runoff and sediment yield per unit area within the system)	Dependent	Independent	Independent
8 Drainage network morphology	Dependent	Dependent	Independent
9 Hillslope morphology	Dependent	Dependent	Independent
10 Hydrology (discharge of water and sediment from system)	Dependent	Dependent	Dependent

ACTIVITY 4.3

Imagine you are standing on top of a high mountain on a clear day, looking out at surrounding peaks and valleys. From your high perch you can see exposed rock, various patches of vegetation (including forest), water courses and glaciers. You wish to explain why the natural landscape you see before you has acquired the topographical characteristics it has. Is this landscape a natural kind? It certainly *contains* natural kinds (e.g. specific kinds of rock, specific species of flora and fauna etc.). But can you be sure that, when aggregated, these natural kinds comprise another one that exists at a larger scale (the landscape scale)? Is it not possible that you are arbitrarily drawing boundaries around what you see, treating it as a discrete landscape when in fact there is nothing 'natural' about the boundaries at all?

If this Activity has left you floundering, then it's because there's much uncertainty about whether physical geographers study natural kinds – an uncertainty that increases the larger the spatio-temporal scale of analysis. Some believe that field sciences like physical geography do not study natural kinds but, rather, the *relationships between* natural kinds. On this view, physical geography in effect contrives its subject matter (see pp. 217–8 when I discuss the issue of 'closure' and 'nominal kinds') because such things as rivers, forest ecosystems and climate are composite phenomena that are 'ontologically fuzzy' – they only exist by virtue of their more fundamental constituents (see Figure 4.6). However, a counterview, linked to the idea of emergence, is that it is the relationships between natural kinds that generate new effects irreducible to those kinds. On this view, physical geographers *do* study 'real' environmental phenomena because these phenomena are 'greater than the sum of their parts' (see Keylock 2003; Rhoads and Thorn 1996).

Third, physical geographers are increasingly questioning whether the biophysical world operates in a regular, consistent, and deterministic way. For several decades the presumption has been that there is an enduring order inherent to bio-, hydro-, litho-, cryo-, pedo- and atmospheric systems. During the 1950s and 1960s, so-called 'functional' studies sought to identify the character and causes of regular spatial patterns within the

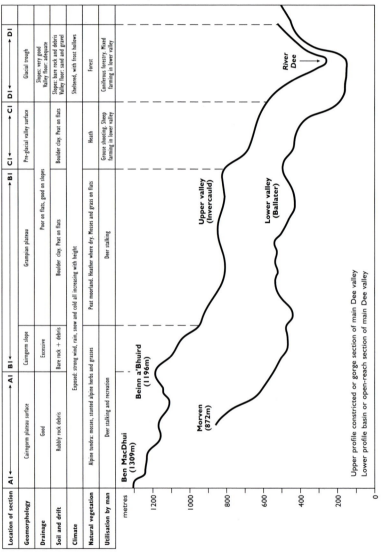

Figure 4.6 Cross-section of a Cairngorm landscape, Scotland. Are you viewing a 'natural kind' or something composed of natural kinds? If the latter then is the cross-section an arbitrary or contrived unit of analysis?

biophysical world (like river meanders and atmospheric depressions). These patterns typically occurred at the meso- or macro-scales. Later, smaller-scale process studies revealed that these patterns were not as regular as previously thought because the processes generating them were affected by various 'intervening factors'. Even so, these process studies still tended to presume a set of fairly consistent (if complex) relationships between environmental processes (like wind and water motion) and environmental patterns and forms (like vegetation height and river profiles). Recently, however, many physical geographers have challenged the 'equilibrium' ideas that have been so popular in their discipline (Box 4.5). In simple terms, equilibrium ideas suggest that all parts of an environmental system are adjusted to the flows of energy and materials that pass through them. Any perturbation of a system's equilibrium state (unless it is very strong) will normally lead to a process of 'negative feedback' that will restore the system to its original state (a process of homeostasis or self-regulation). However, equilibrium ideas are now being challenged by those derived from chaos theory, complexity theory, quantum mechanics and the so-called 'new ecology'.[12] The precursor to these ideas in physical geography were those concerning environment thresholds (Brunsden and Thornes 1979), inspired in part by catastrophe theory (Graf 1979). These various non-equilibrium ideas contend that the biophysical world behaves in irregular, inconsistent and non-deterministic ways. They suggest that it can 'switch' between stable and unstable behaviour depending upon the circumstances (see Phillips 1999). In short, non-equilibrium thinking challenges the idea that there is, ontologically speaking, an inherent 'balance' between the various elements of the non-human world. Among other things, this has profound implications for environmental management. For instance, if the same set of human actions on an environment can have radically different effects depending on the precise environmental conditions, then management measures must adapt to this possibility rather than being of a blanket nature. Phillips (1999) provides a concrete example. He analyses the response of hardwood swamps on the Atlantic and Gulf coasts of the USA to external forcing (e.g. sea-level change or alterations of sediment inputs due to human influence). He shows that even small alterations in forcing variables can produce divergent responses within and between hardwood swamps. So rather than responding in a uniform way, Phillips shows that the behaviour of swamps varies over space and time – in some cases remaining stable, in others tending towards drying out or submergence.

Box 4.5 EQUILIBRIUM ONTOLOGIES

Until fairly recently, the belief that biophysical systems tended towards 'equilibrium states' was common in physical geography. In simple and very general terms, the equilibrium idea proposes that the various components of biophysical systems adjust to one another over time so that they form a relatively stable relationship. Indeed, this relationship is considered to be so stable for many biophysical systems that any disturbance or external 'forcing' will ultimately be compensated for as the system trends back to an equilibrium state (a so-called 'homeostatic reponse'). For instance, in geomorphology it was a long-standing truism that landforms evolved over time in response to dominant environmental processes. Likewise, biogeographers assumed for many decades that 'climax communities' were the norm in the plant world. Climax communities are those assemblages of plant life that are most fully adapted to the prevailing environmental conditions (like climate), albeit modified by local variations in soils, relief etc. In practice, physical geographers have utilised a range of equilibrium ideas in their research. But we can make a broad distinction between ideas of static (or steady-state) equilibrium and more dynamic (or slow-changing) equilibrium. From the late 1970s, geomorphologists in particular began to modify and challenge these ideas. For instance, Schumm (1979) identified intrinsic and extrinsic thresholds. Thresholds are points in which an environmental system undergoes a sudden change, without there necessarily being any alteration in the flows of matter and energy entering and leaving the system. After this sudden change the system may attain a new equilibrium. Subsequent to the threshold idea taking hold in parts of physical geography, complexity theory, chaos theory, quantum mechanics and the so-called 'new ecology' have been drawn upon to inform physical geographers' ontological assumptions. These days it is accepted that many environmental systems do not conform to equilibrium behaviour. Equilibrium and post-equilibrium thinking in physical geography speaks directly to the third definition of the term 'nature' discussed in Chapter 1: that is, nature as a 'inherent force'. The debate is over what form that 'inherent force' takes: order (equilibrium) or disorder (chaos, complexity). A good discussion of the equilibrium idea in physical geography can be found in Inkpen (2004: ch. 7).

Finally, related to the debate over complexity and divergence is one about the balance of general and particular factors in explaining environmental phenomena. As a field discipline, physical geography studies both 'immanent' processes and 'configurational' factors (Simpson 1963). In other words, it examines processes that might be general and universal (like those specified by the laws of Newtonian physics or thermodynamics). But it *also* examines how these processes operate 'on the ground' both together and in conjunction with phenomena they are responsible for (like landforms or weather systems). As part of what Massey (1999) calls its 'physics envy', physical geography has arguably long been fixated on identifying the general processes giving rise to specific environmental phenomena – a fixation going back to the 'spatial science revolution' of the 1950s and 1960s (see Chapter 2). Recently, though, the balance has swung more towards a concern with the importance of the configurational. The argument has been that so-called 'universal' processes cannot be abstracted from the specific circumstances in which they operate. That is, the 'initial conditions' in which a general process unfolds are seen to have a profound influence on the effects of that process in the landscape. For instance, the functional studies of fluvial geomorphologists typically showed a lot of 'scatter' or 'noise' that was not explained by the general theories, models and laws deployed. However, from the 1980s, this scatter was not seen as 'deviant' but as important in its own right as an index of the specificity (even uniqueness) of the phenomenon under investigation. Thus reach-scale studies were replaced by smaller-scale, in-depth investigations of specific river bends, rapids, confluences, and the like. These investigations suggested that the biophysical world is more differentiated than physical geographers had supposed. General processes (like gravity and energy conservation) were shown to have different effects in different times and places – particularly at the small scale, where the specifics of river bed or bank-form, for instance, could alter the operation of processes (like turbulence) governed by general laws of mechanics (see Lane and Roy 2003). In sum, physical geographers are currently debating whether theirs is ultimately a *nomothetic* field science (concerned with general processes underpinning various patterns and forms) or ultimately an *idiographic* field science (concerned with unique patterns and forms underpinned by equally unique conjunctions of general processes with specific local conditions).

Epistemological issues

All ontological beliefs depend, in part, upon epistemological ones. What we believe to *be* real is influenced by how we think we can come to *know* reality. Here I identify three important epistemological issues that physical geographers have wrestled with in recent times. The first is the question of whether or not 'seeing is believing' (what's usually called empiricism). Increasingly, physical geographers argue that there is more to reality than meets the eye. Though physical geography is heavily empirical this does not mean it is necessarily empiricist. Fluvial geomorphologists influenced by the philosophy of transcendental (or critical) realism have made this argument very forcefully (see Richards 1990). Transcendental realists argue that we need a 'depth ontology' if we're to understand reality properly. A flat, empiricist ontology implies that reality consists only of what is observable (and is an ontology characteristic of positivism). A depth ontology, by contrast, makes a distinction between structures, mechanisms and events. Structures are invisible but real elements of reality (like gravity or energy conservation) that undergird the behaviour of many phenomena. In turn, these structures operate on animate and inanimate matter and are expressed as mechanisms (like water turbulence or air convection). Finally, these mechanisms give rise to visible effects (events) of the kind that physical geographers study. For transcendental realists this depth ontology undermines empiricism and ensures that the researcher plays an active role in identifying what is real as opposed to merely observable. For instance, a fluvial geomorphologist studying a pool-and-riffle sequence may surmise that multiple structures and mechanisms intersect to create the phenomena observed (see Figure 4.7). What is more, this conjunction of causal processes may be influenced by so-called 'contingent conditions' (like the morphology of the river bed and profile) such that the same conjunction would produce a different pool-and-riffle sequence elsewhere. What this means, then, is that the researcher must use experience, logic, creativity and imagination to identify the various causes of the phenomena in question – what's sometimes called abduction (see Box 4.6). Since causes rarely operate in isolation for transcendental realists, then there is no one-to-one relation between cause and effect (see Figure 4.8). Instead, the physical geographer is confronted with both equifinality (where different processes can lead to the same outcomes) and multifinality (where the same processes can produce different outcomes). This means that an empiricist approach is an inadequate epistemological basis for deriving knowledge of biophysical reality.

Second, physical geographers have been sensitive to the way they actively bound or enclose their objects of analysis. The 'problem' of closure can be stated as follows: when a physical geographer studies an aspect of the biophysical world they inevitably places boundaries around it episte-mologically and methodologically. For instance, an arid geomorphologist

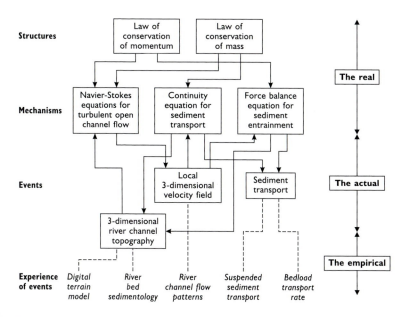

Figure 4.7 A transcendental-realist conception of the relationship between the real, the actual and the empirical in fluvial geomorphology. Reproduced from Lane (2001)

Box 4.6 ABDUCTION

Abduction is a process of working back from an observed effect to a possible cause or causes. It is an imaginative act of conjecturing what would have caused the effect to occur, even though the cause (or causes) cannot be observed and cannot be identified for sure. Causal processes are often invisible and they interact with other phenomena in such a way that their existence is often not easy to identify. Abduction is central to critical (or transcendental) realist thinking and especially necessary in physical-geography research. Since many of the phenomena studied by physical geographers are polygenetic – having multiple causes operating simultaneously – it is not at all easy to work back from effects to causes. This is particularly true at large spatio-temporal scales – as when one is trying to explain landform evolution in, say, the Andes. This is where careful abduction comes in. Abductive reasoning involves disentangling in the mind causal mechanisms that, in reality, might interact differently in different situations to produce the same (or different) visible effects. The researcher wields a mental scalpel, as it were, cutting into the connective tissue of the world at different angles in order to identify the constituent factors at work. Thus the major processes involved in creating and shaping the Andean mountain range might have different effects in the Himalayas. But this can only be established by reasoned and rigorous abduction from visible evidence to possible causal mechanisms. Using existing knowledge of likely causal mechanisms one conjectures as to whether and how they are operative in the Andean case. Subsequently, new evidence can be analysed that may confirm or disconfirm the explanation abductive reasoning produces (and so on iteratively in a virtuous spiral of conjecture and empirical testing).

---

studying an arroyo severs their object of study from its wider local and regional context in order to focus in depth on gully formation. The question then arises: is the geographer in question making the right 'cuts'? Can one arroyo be studied without reference to its connections with other

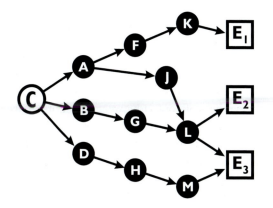

Figure 4.8 The complex relationship between causes and effects. Here the same cause can have either different effects or similar effects depending on the intervening variables (black circles). Adapted from Inkpen (2004)

elements of the surrounding landscape? As soon as the investigator decides on the object of their analysis they have *already* presupposed that it can be studied as an object – one relatively autonomous from other objects or other scales of analysis. This introduces the possibility that physical geographers are putting false boundaries around phenomena whose proper analysis require a different scale of investigation (see Lane 2001: 249–53; Church 1996).

Finally, the problem of closure shades into the debate over whether physical geographers study so-called 'nominal kinds'. Nominal kinds are the opposite of the natural kinds discussed in the previous subsection. They are creations of the analyst not 'real' things. For instance, one might argue that the Canadian shield is a nominal kind. Though it undoubtedly exists, it is arguably a composite phenomena that consists of multiple processes, landforms, water courses, soil types, vegetative communities and so on, interacting over wide spans of time and space. It is thus not a thing 'in itself' and nor can it be readily separated from global biophysical systems like the atmosphere. In this sense, some have argued that physical geographers in effect *create* their objects of analysis. For instance, Vic Baker (1999) has argued that he and his colleagues are really pragmatists (or idealists). Pragmatism, in the sense Baker means it, is a philosophical approach similar in spirit to that animating this book. It argues that we use words, concepts and images to create understandings of a world that is, paradoxically, unintelligible outside those words, concepts and images (see Inkpen 2004: ch. 2). Baker's contentious arguments challenge the realist beliefs of most physical geographers and are similar in their thrust to Demeritt's ideas and those of Feyerabend. More generally, they resonate with the 'discursive construction of nature' ideas discussed in the previous chapter. As such, they offer one of the few points of contact between the de-naturalising tenor of human-geography research into nature and the community of physical geographers. Yet one suspects that most physical geographers see Baker as a maverick not to be take too seriously. As with Demeritt's research, Baker's is doubtless seen as too threatening to physical geographers' conventional self-understanding.

A DIVIDED DISCIPLINE

Let me summarise what we've learnt in this chapter. The ontological and epistemological debates discussed above show that physical geographers are by no means naïve realists. Likewise, the discussion of method in the section 'Producing realistic environmental knowledge' gave some sense of how reflexive (or self-critical) physical geographers are regarding their investigative procedures. Yet for all this, these geographers mostly take it as given that the biophysical world is a real and objectively knowable entity. Whether studying 'natural' or modified environments, they see their role as producing accurate knowledge about the earth's surface. As we saw in the section 'Environmental realism' their willingness to use the term 'science' to describe the process and products of their research is indicative of this fact. Added to this, we saw in the section 'The social construction of scientific knowledge' that physical geographers have generally ignored or deflected attempts to question the possibility that accurate knowledge of the biophysical world is, ultimately, an achievable goal. This is doubtless why the interesting ontological debates within the physical geographic community revolve mostly around how the biophysical world is structured rather than whether it exists as a realm in its own right.

It would be easy to overstate the differences between physical and human geographers regarding the knowledges of nature they produce. Insensitively used, labels like 'realism' and 'social constructionism' can suggest that physical geographers have nothing whatsoever in common with their human counterparts. Clearly, there are major differences of approach. But these should not be exaggerated or misrepresented. For instance, the human geographers whose work I discussed in Chapter 3 are all realists in the obvious sense that their claims about what we call nature are, in their view, not just made up. Similarly, we've seen in this chapter that, while they make use of a different language to most human geographers (that of 'science'), physical geographers do not have a simple-minded belief that the environment can 'speak for itself' if studied in the 'correct' way. In light of this, see if you can answer the following Activity question.

ACTIVITY 4.4

Think carefully about what you've learnt in this and the preceding chapter. What, in your view, are the main points of difference between human and physical geographers vis-à-vis 'nature'?

The answer to this question seems to me to be threefold. First, as mentioned earlier, human geographers are preoccupied with societal representations of nature (including the human mind and body), as well as the processes whereby nature is rendered materially 'unnatural'. Physical geographers, meanwhile, are preoccupied with specifying the biophysical properties of the non-human world, whether it be humanly modified or not. Second, human geographers – despite the epistemologically and ontologically realist pretensions of their own 'nature-sceptical' claims – take a broadly constructionist approach to nature in both the epistemological and ontological sense. Physical geographers, by contrast, ultimately hold on to the idea that the environmental knowledge they produce is, however provisionally, the best and most accurate account we have of the biophysical world 'out there'. Third, and finally, physical geographers produce cognitive knowledge of the environment for the most part. In keeping with their self-identity as 'scientists', they normally separate questions of 'fact' from questions of value.[13] The critical human geographers I discussed in Chapter 3, by contrast, routinely pass judgement on the way those things we call 'natural' are either categorised or physically used/changed. The de-naturalising thrust of their research is intended to call into question certain societal representations, processes and practices.

Well over a century since the 'geographical experiment' was inaugurated what are the implications of this state of affairs? First, the discipline of geography is no longer characterised by the quest to bring nature and society under one explanatory umbrella. These days, only a few environmental geographers and a handful of applied physical geographers study the material interactions between the human and non-human world in any detail – and virtually no one in the discipline entertains ideas about people's mental and physiological nature being either fixed or determining their behaviour and worth. Second, the knowledges about nature produced by geographers fall broadly into two types. Third, these nature-sceptical and nature-endorsing knowledges are derived by geographers drawing upon different repertoires of methods, theories and philosophies.

In short, the topic of nature remains a problem for geographers just as it has done throughout the discipline's history. Not only is there is no disciplinary consensus on what nature is and how to study it, but this lack of consensus also holds human and physical geographers apart – so much so that few in the discipline occupy the middle ground where the two sides of geography meet. My own view is that this is no bad thing and is thus not a 'problem' at all. Some geographers lament the estrangement of human and physical geographers. They worry that geography lacks intellectual integrity at a time when it should be unifying around the study of the human impact on the environment (see, for example, Liverman 1999). The counter-argument – one that I'd endorse – is that the presence of nature-sceptical and nature-endorsing perspectives within one disciplinary space is intellectually healthy. When it comes to nature, geography is an indisciplined discipline – one lacking a 'party line'. This and the previous chapter have, I hope, revealed the strengths (and weaknesses) of claiming that nature is not natural and of claiming, by contrast, that what we call nature is knowable in its own right. The important thing is to avoid taking any claims about nature – whether phrased in a social constructionist or more naturalistic mode – at face value. As I've argued throughout this book, we need to examine what motivates geographers to insist that nature either is, or is not, what it appears to be.

EXERCISES

- Make a list of all the reasons why you, as a student, would normally believe that the environmental knowledge produced by your physical geography professors is truthful and accurate. In light of what you've learnt in this chapter, how many of these reasons do you think stand up to close scrutiny?

• List some of dangers that arise if we apply the nature-sceptical sensibilities of many human geographers to the environmental knowledges produced by physical geographers.

FURTHER READING

Two excellent introductions to science have been written by Woolgar (1988) and Sardar and van Loon (2002). Good primers on the philosophy and methods of science are provided by Chalmers (1990) and Okasha (2002). Despite its focus on social science, Smith's (2002) book offers lucid discussions of many of the issues covered in this chapter. Kukla (2000) confronts the social constructivist challenge to scientific realism head on. In physical geography, the best general texts on scientific method are Haines-Young and Petch (1986), Inkpen (2004: chs 1-5) and Schumm (1991), while Marshall (1985) offers a good account of scientific method for geographers more generally. Rhoads and Thorn (1994) set these methodological debates in a wider intellectual context. The overlapping ontological and epistemological debates discussed in this chapter are well discussed by Burt (2005), Inkpen (2004: ch. 7), Lane (2001), Phillips (1999), and Thorne (2003). These debates have also come to the fore in responses to papers by Massey (1999) and Harrison and Dunham (1998) – see the *Transactions of the Institute of British Geographers* (1999 [24] 2; and 2001 [26] 2). A debate on method in physical geography in the *Annals of the Association of American Geographers* (Bauer 1999) is also worth referring to.

In a sense, the ontological and epistemological debates current in physical geography indicate that the non-human world is a 'problem' for the discipline. Another indication of this problem is the splintered character of physical geography – a result of the fact that specialisation has been deemed necessary in order to understand the huge range of environmental phenomena that physical geographers confront. Against this, some have argued that global environmental problems offer an opportunity for physical geographers to focus again on interconnections between the atmo-, pedo-, hydro-, cryo- and biospheres. See Gregory et al. (2002), Gregory (2004) and Slaymaker and Spencer (1998: ch. 1) for viewpoints.

I argued in the chapter that physical geographers have resisted constructivist criticisms of the knowledge they produce. Symptomatic of this is the way the ideas of Thomas Kuhn have been used. Kuhn coined the term 'paradigm' (see Box 1.6) and argued that sciences progress through 'revolutions in thought' not a steady accumulation of knowledge. Geography as a whole fixated (during the 1980s) on the issue of whether Kuhn's idea of scientific revolutions best describes intellectual change in the discipline. But Kuhn's more interesting contribution was to challenge the realist credentials of science. The notion of paradigm incommensurability suggested that different researchers in effect saw a different world because their paradigms so conditioned how reality is apprehended. In physical geography, Haines-Young and Petch (1986: ch. 4) and Sherman (1996) are among the very view to take seriously this dimension of Kuhn's thinking. Bassett (1999) deals with the more general issue of whether there is academic 'progress' in geography over time.

On the whole, physical geographers see themselves as producers of cognitive knowledge about the biophysical world. A few, however, have considered if and how values can enter their work: see note 13. In the final chapter of his book, Inkpen (2004) reflects upon the social networks physical geographers move in and the way these affect the what and how of their research.

The intellectual diversity and disunity of geography has long between a debating point within the discipline. Human and physical geographers' differing understandings of nature are just one reason for the divide between them – but how important a reason? For an answer to this question see Urban and Rhoads (2003), Viles (2004) and Castree (2001b).

5

AFTER NATURE

'it's terribly important to overcome these divides . . . and it's terribly hard
to find a language to do so'.

(Harvey and Haraway 1995: 515)

INTRODUCTION

In this final, relatively short chapter, I present the ideas of geographers who
might be described as 'after-' or 'post-natural' in their outlook. This growing
cohort of geographers take issue with the society–nature dualism that
underpins most of the thinking reviewed in the previous two chapters.
Deeply ingrained in Western thought, this dualism leads us to divide the
world ontologically into halves. Even though these halves are connected,
we tend to think of them as different. Thus, in Chapter 3, we saw that social
representations and forces were the key to understanding nature for many
geographers. Meanwhile, in Chapter 4, we saw how physical geographers
focus on environmental 'realities' that, in their view, are irreducible to
people's discursive or material practices even though they may be affected
by them. Despite the different approaches to nature discussed in the
previous two chapters they arguably have one thing in common. They
either emphasise a set of phenomena classified as 'social' *or* they emphasise
a set of phenomena classified as 'natural' (whether 'first natural' or 'second
natural'). In each case, one of two domains supposedly comprising our total

reality is given prominence. Accordingly, we can say that an *ontological schism* runs through contemporary geography that few have been inclined to challenge.

The geographers whose work I discuss in this chapter seek to overcome this schism. Among the several reasons for this, three stand out. First, it is arguable that the world is seamless. Dualistic ontologies imagine a world split in two: a world of separate spheres that 'interact' and 'collide' or where the character of one is 'constructed' and 'determined' by the other. But this risks severing the ties that, in reality, make the two plutative 'spheres' indissociable. Second, it is arguable that the ontological differences *within* the 'social' and 'natural' worlds are as large as those said to firmly distinguish them from one another. For instance, we might ask why are a rock and a gorilla thought to have more in common than a gorilla and a human being, when the latter two are both primates? Third, it is arguable that the society–nature dualism blinds us to the need for a new vocabulary to describe the world we inhabit. This would not be a vocabulary of 'pure forms' but one that captures the hybrid, chimeric, mixed-up world in which we are embedded.

For these reasons (and others to be explained), a strong minority of geographers eschew the nature–society dualism that most of us regard as so normal to be common sense. These geographers beat a 'third way' between this and related dualisms (like subject–object, urban–rural, and people–environment). They do not, for instance, talk about a socially con-structed nature because they resist the idea that 'society' is a self-sufficient domain that can 'construct' something external to it. Likewise, they do not talk about the 'non-human world' or 'human biology' because this implies they are discrete domains with equally discrete properties or else surfaces upon which society inscribes its wishes. The geographers whose work I discuss in this chapter are thus best described as *relational* thinkers. Relational thinkers argue that phenomena do not have properties in them-selves but only by virtue of their relationships with other phenomena. These relations are thus *internal* not external, because the notion of external relations suggests that phenomena are constituted *prior to* the relationships into which they enter.

This probably sounds as abstract as some of the ontological and episte-mological debates recounted in the previous chapter. But its implications are highly concrete. As we'll discover in this chapter, a cohort of 'after-' or 'post-natural' geographers wish to alert us to a world existing under our noses but one which we fail (quite literally) to see if we divide it into natural and social things (or some combination of the two). These geographers are not all of a piece – for instance, they include so-called 'actor-network theorists' and certain Marxists, two groups who disagree on many things. But some broad commonalities none the less underpin their research. Intellectually, the project of these geographers is to reveal a more-than-social, less-than-natural world to us and so challenge what they see as the impoverished vocabulary we use to analyse it. Morally, their project is to move us away from the ethical codes that ground their claims in natural imperatives or socially contingent assessments of what we call nature. Some readers will find the material discussed in this chapter disconcerting because it refuses the mindset of the work reviewed in the previous two chapters. As before, my aim to ask why some geographers want us to see nature in the ways explored in this chapter. What is their aim in arguing that we are 'after nature' in so far as nature, in their view, does not exist either as a social construction or as a realm irreducible to social representations and forces?

Before proceeding I should enter two points of clarification. First, many of the geographers whose work I discuss below take issue with the framework of analysis used in this book. This framework leads me to treat the ideas of these geographers as just that: ideas about the world vying with others for our attention. As we will see, some of these geographers dispute the notion that we re-present a world 'out there' in knowledge. This is because they reject the subject–object dualism that apparently animates my framework – a dualism that posits a world of things-in-themselves on the one side, and our knowledge of those things in speech, writing and imagery on the other. Second, it is important to avoid a fundamental error that is all too easy to make when presenting the work of 'after-' or 'post-natural' geographers. This is the error of supposing that where the society–nature dualism *was* appropriate until recently it is *now* obsolete because 'technoscience' has breached the ontological divide between society and nature. Against this 'epochal' error, the geographers whose work I discuss below maintain that we have *always* lived in a mixed-up, hybrid and 'impure' world where it is difficult to disentangle things from their relationships. Technoscientific developments like transgenic pigs, smart robots and microchip implants are, in these geographers' estimation, just the latest examples of a long history of society–nature interfusions.

The chapter begins with an example of how, generally speaking, a 'post-natural' analyst views the world, after which I discuss the main lines of a non-dualistic, relational mode of thinking in contemporary geography. Thereafter, I focus on the work of Sarah Whatmore and inquire into the moral implications of the seamless ontology being advocated. This leads to some speculations on what motivates geographers like Whatmore to make the arguments they do.

NEITHER NATURAL NOR SOCIAL

So ingrained is the tendency to employ the society–nature dualism in our Western way of thinking that we forget that it's anything but natural. Not only do many non-Western societies not carve the world in two in this way, if we look back historically we also see that Westerners themselves only began to employ the dichotomy from the eighteenth century, during the so-called Enlightenment period.

The nature–society dualism implies the mutual exclusivity of its two sides. So long as we operate with this dualism we are forced to concede that (i) society and what we call nature are different and can be studied separately, and that (ii) the study of society–nature relations must resort to notions like 'construction', 'interaction' and 'interrelation' (see Figure 2.1 again). But what would it mean to do away with the society–nature dualism? What would it mean to question the idea of a non-natural society and a material nature that is autonomous from (or else a product of) social representations and forces?

We can begin to answer these questions by revisiting the second of the seven vignettes presented in Chapter 1. This was the story about Britain's 'rainforest', a biodiverse brownfield site that is currently slated for development as a light industry park.

ACTIVITY 5.1

Read this story on pp. 2–3 once again. In light of what you've learnt in the previous two chapters jot down (i) how a human geographer might explain the 'nature' in question, and (ii) how a physical geographer might approach the issue.

If we employ the distinction (made in Chapter 3) between representational and material constructionism we can propose the following interpretation of the story from a 'typical' critical human geographers' perspective. First, we might suggest that the idea of 'biodiversity' is a value-laden construction being used by Buglife and English Nature. This idea, as explained in Box 1.8, purports to merely describe species diversity whereas in reality it arguably puts a positive 'spin' on such diversity as 'inherently good'. As analysts of representation we would thus wish to examine the values and interests of Buglife and English Nature – why, we might ask, do they depict species diversity in the ways they do? Second, a less discursive analysis might argue that the brownfield site is a physical construction – albeit an unintended one. A 'weak' constructionist argument might suggest that the nature found at the disused Occidental facility is not natural because it is the result of human interference. Yet, unlike hybrid crops and GM foods (say), it is not exactly 'produced' because it was not designed consciously by social actors and organisations.

How would a physical geographer (in this case a biogeographer) approach the nature found on Canvey Island? Arguably, they would acknowledge the anthropogenic influence, but they would also insist that skylarks, shrill carder bees, badgers and the like have their own characteristics and modes of behaviour that mark them off from any particular human representations and practices. They would likely focus on the interrelations among species, and bracket off consideration of how the site has been used by children since its abandonment in the early 1970s. Finally, they would have faith in the possibility of producing accurate, value-free knowledge of the site's biodiversity – knowledge that could subsequently be used to inform decisions about whether or not to develop it for industry.

So far so good. But we can propose another way of looking at the Canvey Island site that does not resort to the ontological dualism between social representations and practices on the one side and an ultimately non-social world on the other. What if we see the site as a *network* rather than a place where two 'spheres' come together? A network metaphor gets us looking at the *indissoluble links* between multiple different phenomena. The actions and effects of one phenomenon are *part of* others in the network. Nothing exists in isolation which is why it is mistaken to separate out different classes of phenomena, like 'social' and 'natural' ones. All there are intimately related 'actants' whose existence and effects depend upon those of all other living and inanimate things in the network, past and present. Unlike 'actors' actants are not free agents. Instead, they are both the cause and consequence of all the other entities connected to them. What is more, no two networks are necessarily the same, which is why it is important to attend to the specific conjunction of phenomena in any given situation.

This may sound very abstract, so let's apply the network ontology to the Canvey Island site. Clearly, the site is not a 'natural ecosystem' since it would not exist in its precise form without the actions of Thames dredgers, the Occidental oil company or children playing. But neither is the site a social construction because the biophysical character and actions of non-human species at some level escapes the intentions of human actors. So what is the site? From a network perspective it is a particular alignment of human and non-human actants. Take any of them away and the site would not be what it is today. Thus the wildlife species at the site undoubtedly exists elsewhere in the world but – and it's a crucial but – they do not exist in the same way or with the same effects on proximate flora and fauna. The silts dumped over former fields and marshes, the disused buildings and equipment, the tracks and trails created by playing children and bikers: these and other interventions have inadvertently created opportunities for a uniquely diverse array of wildlife to not merely co-exist but co-depend. The 1,300 species at the site are, in various ways, reliant upon each other for survival, just as they have afforded opportunities for tactile play and enjoyment for human actors over the years by virtue of their material properties. In reality, the various actants are so thoroughly stitched together that it is arbitrary to group them into two major categories, imagining that the domains so categorised 'come together' like two separate pieces of a jigsaw. In sum, then, a network approach to the world is 'post-natural' because it eschews big ontological categories for a micro-level focus on the specific actors and the relations between them that constitute our world.

THINKING RELATIONALLY

My use of a network metaphor above speaks to one of the four major variants of post-natural thinking current in geography ('actor-network theory', which I will discuss below). This thinking has been advocated by human geographers for the most part. Interestingly, physical geographers have not made more of the potential for overcoming the society–nature dualism latent within complexity theory, chaos theory and affiliated ontological positions on the biophysical world. In the main, physical geographers adhering to these positions limit their reach to environmental systems and bracket out the human influence for analytical convenience. Yet ideas about chaos and complexity could, in principle, form a unifying ontology for geography as a whole – rather as systems theory was intended to bridge the human–physical divide in the 1960s and 1970s. Likewise, the ideas discussed below could bridge that divide – but only if enough geographers on both sides of geography buy into them. Needless to say, the potted summaries of the four main strands of relational, 'post-natural' thinking I identify do not do justice to any of them.

Non-representational theory/performativity

Non-representational theory is most closely associated with the work of the British human geographer Nigel Thrift. The word 'theory' is rather inappropriate here. Thrift has laid out a set of general principles and arguments about representation and alternatives to it rather than a specific theory about how society and nature intertwine. To understand Thrift's complaints about representation it is useful to re-read the prefatory comments I made in the section 'Re-presenting nature' of Chapter 3. There I observed that, according to analysts of representations of human and non-human 'nature', there are two elements to consider whatever the specific representation in question (visual, written or verbal). First, there is an act of 'speaking for' – where the represener acts as a representative of that which is represented. Second, there is a simultaneous and less obvious act of 'speaking of' – where the represener actively 'frames' the represented while claiming to re-present it 'as it really is'. Both elements of representation were exposed in Braun's study of Clayoquot Sound. The point that Braun and other critics of representations of nature make, you will recall, is that representations 'construct' the realities they purport merely to depict.

Thrift (1996; 2003) makes a number of criticisms of the preoccupation of many human geographers with representations of nature (and other things too). First, he argues that this preoccupation wrongly implies that people relate to the material world primarily in visual (or ocular) terms, leading to pictorial, written and spoken representations of it. Thrift reminds us that we engage with the material world (including our own bodies) using *all* our senses: we are practical beings not just intellectual ones. Much of our understanding of, and action upon, this world is thus, in Thrift's view, never formally represented or representable at all because it is tacit, sensuous, habitual and precognitive. Second, Thrift argues that a focus on representation wrongly implies that we humans are *distanced* from the material world – as subjects viewing objects – rather than beings who *inhabit* this world. Finally, Thrift argues that it is wrong to ask whether representations of what we call 'nature' contain the hidden agendas of representers or some 'truths' about that which is represented. For him, representations are one of several *tools* we use to make sense of the world. Representations of reality help us to make our way in this world, not because they are 'right' or 'wrong' but because they have effects on how we act to the extent that we accept them. What's more, in Thrift's view, all representations arise from and affect our practical *engagements* with the biophysical world, be it 'natural' or humanly altered.

In sum, Thrift holds to a worldview that does not separate knowing (epistemology) from that which is known (ontology). His is a non-representational approach that focuses on a world in which we are dwellers not observers, multi-sensual participants not detached spectators. We come to know by doing, and we do because of what we already know in an iterative process where the material world affects us and we affect it. In Thrift's seamless conception of reality, a society–nature dualism is thus too crude to be of use. For him, we are not 'minds in a vat' whose representations of nature reflect our self-sufficient values, aspirations and biases. Rather, we are constitutent parts of a 'more than human' world without which we could not become the sorts of sophisticated thinking and acting beings we become over our life-course. Thrift thus ultimately sees the world not as a pre-existing collection of human and non-human entities but as a set of mutually constitutive encounters or performances. His project is to reinject some 'life' into the lifeless landscapes revealed by analysts of lay and expert representations. Thrift's work has rapidly created a 'school' of non-representational thinking, one carried forward in the UK by his former graduate students for the most part. This school is now beginning to apply Thrift's general ideas to specific empirical contexts.

Actor-network theory

Actor-network theory (ANT) has become closely associated with the sociologists and anthropologists of science Bruno Latour, Michel Callon and John Law. It has, of late, been as influential in British human geography as non-representational theory, and is also now having a wider influence in anglophone human geography. Jonathan Murdoch, of Cardiff University, has been an effective, but not uncritical, populariser of ANT thinking (see Murdoch 1997a; 1997b). Like non-representational theory, ANT is not a 'theory' in the strict sense of the term. Instead, it's a set of overlapping propositions intended to alter conventional thought and research regarding the relationships between those things we routinely think of as 'social' and 'natural' respectively. Fundamentally, it challenges the 'two spheres' assumption underpinning the discursive and material versions of social constructionism discussed in Chapter 3, as well as the 'natural realism' discussed in the previous chapter.

In the first place, ANT suggests that the society–nature dualism illicitly simplifies a world that is much *messier* than we allow. This world does not divide neatly into two ontological domains but is, rather, characterised by myriad qualitatively different but intimately related phenomena. Second, ANT makes much of the network metaphor I used in the previous section of this chapter. It sees the world as consisting of multiple, cross-hatching networks: that is, assemblages of human and non-human things that are aligned in more or less ordered ways. For instance, rather than seeing OncoMouse – a genetically modified kind of mouse used in cancer-drug experiments – as a 'material construction' of scientists and pharmaceutical firms, we can see it as part of a network. Its existence in the particular 'unnatural' form it assumes is not *simply* attributable to intentional human actions. It also depends upon a whole array of highly specific non-human instruments, including sophisticated laboratory equipment, scientific papers containing information on how to modify mice genetically, and electronic flows of money to fund OncoMouse's ongoing production. These non-human instruments are utterly essential for human intentions to be realised in this case: they are indispensable 'intermediaries' that connect OncoMouse to its human originators. Third, this brief example allows us to see why ANT talks about *actor-networks* rather than networks alone. Networks of human and non-human phenomena are neither more nor less than their constituent parts in ANT. Each part has whatever role (agency) it has not only by virtue of its intrinsic properties but also because of its position *relative to* other agents in the network. This is why ANT uses the neologism 'actant' rather than the conventional terms actor or agent to describe the material role of human or non-human phenomena.

Plate 5 Genetically modified mouse with 'natural' sibling. Is the GM mouse 'unnatural', a social construction or neither? Is it, perhaps, an actant in a network each of whose constituents has a role to play so that categories like 'natural' and 'social' begin to look inadequate? (© Science Photo Library)

On the basis of its 'stringy' ontology, ANT makes no assumptions about which actants in any given actor-network are marshalling all the others. It thus refuses the *a priori* choices of social constructionism and 'natural realism' recounted in the previous two chapters. ANT researchers prefer to pay close empirical attention to actor-networks in their specificity, showing how human and non-human phenomena co-constitute one another in any given case. As with Thrift's work, ANT aims to respect the intimate weave of life without talking about an asocial nature (human or non-human) or a non-natural society. ANT insists that we have *never not* lived in a 'hybrid' world where what we call 'social' and 'natural' things are so closely entwined that these labels make little sense.

New dialectics

Where ANT makes no assumptions about what, if anything, different 'socio-natural' networks have in common, so-called 'new dialecticians' look for overarching *processes* that structure these myriad networks. David Harvey is the most vocal exponent of this process perspective, notably in his book *Justice, Nature and the Geography of Difference* (1996). This perspective takes issue with the Cartesian, Newtonian and neo-Kantian worldviews that, in Harvey's view, dominate Western thought. These worldviews imagine the human and non-human worlds to be composed of discrete physical *things* which can be analysed prior to, and separate from any contingent relationships they may have with other things. This *atomistic* perspective (see Box 4.3) sees the relations between human and non-human phenomena as external ones that play no necessary role in constituting those phenomena. Against this, Harvey adopts what I earlier called an *internal relations* perspective, following the Marxist philosopher Bertell Ollman (1993) and the Marxist biologists Levins and Lewontin (1985). In this perspective, what makes a given thing different from and apparently unrelated to other things is, in fact, its relations with those things. These relations are, as it were, part of or contained within the phenomena that are related to one another in any given case.

This links directly to the notion of process mentioned above. Process involves change in one or other direction. It involves linking diverse phenomena in order that certain goals or ends are achieved, whether by accident or design. As a Marxist, the key process that Harvey uses to illustrate his internal relations perspective is capitalism. This may sound strange at first hearing. After all, capitalism is an economic system. So why describe it as a process? A second look at the diagrammatic representation of capitalism in the section in Chapter 3 called 'Re-making nature' provides an answer. We see there that capitalism is about the *circulation* of commodities (e.g. goods and money) and the *expansion* of wealth (in the form of profit). For Harvey, then, it is this overarching compulsion to 'accumulate for accumulation's sake' that links all manner of human and non-human phenomena in intimate ways. From factory-farmed chicken to global warming, Harvey argues that people (e.g. as wage workers) and non-humans become the 'arteries' through which an invisible process of ceaseless value expansion flows. Because this process is seamless, Harvey argues, we should not make the mistake of fixating on the different things that become embroiled in it. While these differences matter, they do so only in relation to the abstract process that conjoins them.

Overall, Harvey thus regards *particular* human and non-human things as the expressions of *general* processes. He terms these things 'moments' that give physical form to the general processes involved. Note that this differs from the idea of a 'social construction' because Harvey insists that the material properties of non-human things used for human purposes have a not-always-predictable role to play. The 'dialectical' element of all this is that from time to time particular moments (physical things) contradict the 'demands' being placed on them by the logic of process. For instance, from Harvey's Marxist perspective, most fish-farming can be seen as an economically rational response to overfishing in the high seas. Farmed fish thus become physical means to realise profits for fishing firms at a time when the 'natural' fishing industry faces a profitability crisis in many parts of the world. But farmed fish are forced to behave in ways rather different from their 'natural' brethren, not least because they are concentrated in large numbers in small pens. This has already led to serious disease outbreaks among fish-farm populations and has been an unintended consequence of fish-farming as a relatively new technological practice. There is thus here a contradiction – an internal one – between the 'logic of capital' and the physical means through which that logic is being expressed in this case. Diseased fish can cut into profits and, if not properly controlled, undermine fish-farming both economically and biologically as an 'unnatural' alternative to open-ocean fishing.

Harvey's 'new' dialectics is more subtle than older Marxist modes of dialectical thinking. It has been taken up by other Marxist geographers, like Erik Swyngedouw, but has thus far proved less influential than non-representational and actor-network theory. In part, this reflects the current unpopularity of Marxism within human geography.

The new ecology

The fourth body of relational thinking about society and nature I want to briefly discuss is called the 'new ecology'. This has been advocated by, and is quite influential among, environmental geographers. One of these is the American Karl Zimmerer, co-editor of *Nature's Geography: New Lessons for Conservation in Developing Countries* (1998). In this book and a set of published essays, Zimmerer (1994; 2000) has proselytised on behalf of the new ecology, which has emerged from the disciplines of biology, zoology and botany over the past twenty years. The 'old' ecology was characterised by two main things. First, it believed that species existed in relatively stable, predictable relationships both with one another and their surrounding biophysical environment. Second, it tended to treat humans either as well-adapted parts of wider ecosystems or else disruptive forces that failed to respect the integrity of those ecosystems. This second element of the old ecology was particularly evident in post-war environmental geography. As explained in Chapter 2, many 'human ecologists' sought to describe and explain the various ways previous and present-day non-industrial societies utilised their different natural environments in a sustainable way. Following Thomas's germinal *Man's Role in Changing the Face of the Earth* (1956), this focus on 'harmonious' human–environment relationships was supplemented with investigations of how 'modern' people were disrupting the equilibrium of the non-human world more and more.

In contrast to all this, the 'new' ecology pioneered by biologist Daniel Botkin (1990) and others makes two counterclaims. First, it challenges the old ecology's equilibrium assumptions and 'accents disequilibria, instability, and even chaotic fluctuations in biophysical environments, both "natural" and human-impacted' (Zimmerer 1994: 108). Second, it follows from this that when people do make large-scale changes to 'natural' communities of plants, animals and insects they are not necessarily 'disrupting' an evolutionary harmony. As Zimmerer argues, this has major implications for geographical research on people–environment relations, as well for environmental management. Except where dealing with 'traditional' societies, the old ecology licensed forms of environmental geography that placed people outside of and in opposition to natural environments. By contrast, the new ecology's challenge to the 'balance of nature' postulate of its predecessor opens a space for environmental geographers – indeed *all* geographers – to regard human actors as always already *part of* complex and changeable biophysical systems. In terms of managing how people use local and non-local environments, the new ecology also challenges the long-standing beliefs that the human alteration of an apparently stable ecosystem is 'bad' and thus conservation must proceed by way of little or no human interference.

In sum, Zimmerer and other geographers influenced by the new ecology are apt to talk about 'nature-society hybrids' rather than two interacting domains or spheres. Resonating with ANT in spirit if not letter, the new ecology enjoins us to see the world as a mesh of multi-scalar and sometimes unstable knottings of people (with their varied outlooks, economic practices etc.), plants, animals, soils, water, forests and much more. Here the job of research is not to judge human actions against some eternal benchmark of stability imposed by the non-human world. Rather, it is to trace the varied ecological impacts of different human actions (distinguished by type, frequency and magnitude) upon the non-human domain, and vice versa. As with the three modes of relational thinking summarised above, it is charter for a more 'joined-up' geography.

Clearly, there are strong family resemblances between non-representational theory, ANT, new dialectics and the new ecology. The following Activity links all this back to the conventional definition of nature laid out in Chapter 1 and so reinforces the ways the four bodies of thinking are 'after-' or 'post-natural'.

ACTIVITY 5.2

From the foregoing summaries, can you identify how the new post-natural thinking in geography takes issue with the three conventional definitions of the term 'nature' laid out in Chapter 1?

The quartet of approaches discussed above challenge all three main definitions of nature identified in Chapter 1: nature as the non-human, nature as the essence of something and nature as an overarching force. In the first case, these approaches all cross the 'social–natural divide' in our thought by arguing that what we call 'social' actors, representations, institutions, and so on, depend thoroughly on the existence and agency of what we call 'natural' phenomena. As such, they are ontologically 'symmetrical' and make few, if any, assumptions about which social and non-human phenomena have (or don't have) the power to influence the others. They mostly take each case on its merits and seek to identify the specific ties that connect and condition the actors in question. Second, the four approaches discussed above are all non-essentialist. For instance, rather than assume that shrill carder bees are the same wherever and whenever they can be found, ANT would suggest that their behaviour and its effects might vary (albeit within limits) depending on the context. Finally, while none of the approaches discussed deny that universal forces like gravity cross-cut the human and non-human worlds, they all take issue with the idea that there is some single transcendental principle that governs how the world works (like equilibrium and balance).

MORALITY AFTER NATURE

Clearly, the post-natural approaches discussed above challenge the conventional descriptive and explanatory habits of most professional geographers. They disbar attempts to isolate out 'social' and 'natural' phenomena, to study them separately or to link them causally in more or less direct ways using motifs like 'interaction' or 'construction'. But it also follows from this that post-natural approaches challenge the moral and ethical habits of most professional geographers. This is, of course, immensely important. Moral claims about a supposedly asocial 'nature', in both academia and the wider world, remain as widespread as they are potent. As we have seen in previous chapters, when considered at the broadest level, these claims fall into three main kinds. First, most physical geographers insist (or imply in their research) that the 'facts' about nature (in their case the non-human world) can and should be kept logically distinct from any moral claims about nature. Second, most critical human geographers insist that this first position is naïve because real-world actors constantly link facts and values in their discourses about nature. James Proctor (2001) provides an example. He analyses how a piece of scientific research on freshwater species in North America (Ricciardi and Rasmussen 1999) was reported in a 'pro-environment' news website, <http://www.enn.com>. The report's first sentence reads: 'Some freshwater species in North America are becoming extinct at a rate as fast or faster than rainforest species, but their plight is largely ignored, according to a recent study out of Canada'. As Proctor observes, the facts of species loss are here reported in a highly value-laden way that is so familiar that it risks being taken for granted. It is implicit in the reporting that species loss is morally unacceptable and should be halted forthwith. This is another example of the 'moral naturalism' explained in Box 3.3. Finally, following on from this, most critical human geographers would argue that we never derive our ethics directly from the facts of nature (human or non-human). Rather, we construct our ethical codes and these codes vary from person to person and society to society for this reason.

Despite the differences between these stances on ethics and nature, the relational approaches I've outlined take issue with all of them. Why is this? One reason is that all the stances seek to ground themselves in one or other ontological domain. For example, both moral naturalism and moral constructivism claim that our ethical stance on 'nature' is mandated either by nature itself or by specific societies and cultures. Another reason is that the three stances offer us the polar choices of moral absolutism and moral arbitrariness. The former is a characteristic of many attempts to ground our ethics in supposedly 'given' facts of nature – a sort of 'nature knows best' ethics which leaves human actors little choice but to obey. In *Nature's Geography: New Lessons for Conservation in Developing Countries* Zimmerer and Young (1998) offer compelling examples of how much is at stake here. Until recently, they show, the environmental conservation 'wisdom' in many developing countries reflected the values of the 'old ecology' and also romantic views of the non-human world exported by colonial administrators and scientists from the late nineteenth century onwards. This 'people versus parks' wisdom justified the frequent removal of peasants, tribes and indigenous communities from their lands in order to conserve 'natural landscapes' that were supposedly threatened by human land-use practices. In contrast to this, moral arbitrariness is a potential weakness of the first and last of the three ethical stances outlined above. Here our moral perspectives on 'nature' are seen to have no firmer basis than that people have *decided* to adopt them, regardless of the physical 'realities' of the nature in question.

What, then, would a 'relational ethics' look like – one where 'nature' is seen neither as a construction of society with no independent moral status nor as a separate domain to which we should extend moral considerability? Thus far, it's probably fair to say that relationally minded geographers have failed to provide concrete answers. On the whole, they have furnished only the philosophical outlines of a post-natural ethics rather than a discussion of substantive moral principles (like justice, rights and obligations). Even so, we get can get a sense of what this different moral universe might look like by examining Sarah Whatmore's reflections on the matter.

Sarah Whatmore is author of *Hybrid Geographies* (2003) and several influential essays on post-natural thinking (e.g. Whatmore 1997). A professor of environmental geography at Oxford University, she has drawn upon ANT, certain strands of feminism, and the philosophy of Deleuze, Guattari and Stengers to map the contours of a relational ethics. How does 'nature' figure in this ethics? It is clearly not, in Whatmore's view, a discrete class of entities (human and non-human) that either do or do not deserve ethical consideration by people. We must, in her view, do away with the moral codes typical of environmentalists (who focus on the non-human world) and bioethicists (who focus on human physiology and psychology). Like

Thrift, she reminds us that people are *embodied* and *corporeal*: we are what we are because of our *ties* (e.g. through the intake and excretion of food) with countless non-human others. What follows from this, in her view, is that it is necessary for any ethics to take account of many or all of the actants in any given 'more than human' network. We have no choice but to recognise that each of us is connected to local and global 'imbroglios' in which all the component 'parts' play a role. It thus makes little sense, Whatmore argues, for us to argue that ethics emerges from or applies to one or other set of entities. Likewise, we go awry ethically if we think that our morals about 'nature' are dictated *to* us (by external facts) or *by* us (in virtue of our self-sufficient beliefs, values and assumptions). In sum, Whatmore advocates an ethics that is 'generous' in making no assumptions as to who or what might deserve ethical considerability in any given situation. This cosmopolitan ethics is intended to equip us with subtle moral skills that refuse the either/or choice of a socially contrived ethics or a naturally dictated one. It's an ethics attuned to mixity, impurity and the realities of a 'companioned world' (for more on Whatmore's arguments see *Antipode* 2005; for an accessible book written in the same vein as *Hybrid Geographies* see Hinchliffe 2006).

WHAT MOTIVATES POST-NATURAL THINKING?

It would be all too easy to infer two things about the post-natural thinking discussed in this chapter. The first, which I queried earlier, is the idea that this thinking is preferable to either social constructionism or natural realism because it reflects the new hybrid 'realities' of phenomena like people with xenotransplanted organs. The second is the idea that because relational thinking is currently considered *de rigueur* and cutting edge by many geographers it must be 'better' than its putative predecessors. Against both these inferences it should now be abundantly clear that I think it better to ask of post-natural thinking: what motivates its advocates to make their claims and what are they seeking to achieve by moving beyond the society–nature dualism?

This question directs our attention towards the interests and ambitions of geographers like Thrift, Zimmerer and Whatmore. It directs our gaze away from the seamless socio-natural realities these geographers purport to represent to us. This is not, of course, to say that these geographers do not think of themselves as in the truth-telling business. As academics, they clearly *do* believe that they are revealing truths that our normal ways of thinking and acting conceal from us, as both geographers and citizens (see Murdoch and Lowe 2003 for an especially clear example). But this is not all they are up to. They are also making conscious intellectual interventions for specific reasons. What might these reasons be? One reason, hardly exclusive to this group of geographers, is the compulsion to innovate intellectually that is part and parcel of Western academia. As David Harvey (1990: 431) has acknowledged, the competitive relationship between individual academics, their departments and their universities makes criticising current intellectual wisdom professionally profitable for those able to do it successfully. Less cynically, we can speculate that what we see here is an attempt to reintegrate a fairly disintegrated discipline and so renew 'the geographical experiment' in a new, productive way that might ultimately benefit most geographers. Third, we can suggest that a new respect for the world's complexity and fluidity is here being expressed. In a challenging essay, Steven Hinchliffe (2001) shows us what is at stake. His analysis of how scientists seek to capture the ontological 'essence' of prions reveals that the presumption that prions have an essence (i.e. that they are 'natural kinds') hindered rather than helped the resolution of the British 'BSE crisis' in the 1990s. If these scientists had been more attuned to the motility of prions then, Hinchliffe argues, the BSE problem would have been dealt with more effectively. Finally, we can conjecture that a genuine moral concern is being expressed here: a concern that our dichotomous way of describing, explaining and judging the world is having bloody consequences for people and non-humans alike. For instance, Zimmerer's critique of 'people-less nature conservation' is precisely an attempt to do justice to displaced communities while attending carefully to the material qualities and moral rights of the biophysical world.

It's worth noting that the relational thought discussed in this chapter bears some apparent resemblances to ontological holism (see Box 4.4 again). Holism has been an important part of environmental ethics outside geography, both in academia (e.g. professional environmental philosophy) and the wider world (e.g. among deep green activists). Its most famous expression is the 'Gaia hypothesis' of the British scientist and environmentalist James Lovelock. According to Lovelock our planet is a huge, highly integrated system that has an inherent tendency towards order among its various systems and subsystems. Some environmentalists have used the Gaia ('mother earth') idea to argue that if humans abuse the planet then they will be ultimately extinguished as a species by the 'blind' mechanisms of biophysical self-regulation that will ensue. None of the relational approaches explored in this chapter are holistic in this super-organic and homeostatic sense. In fact, all of them would *oppose* this kind of holism on the grounds that it potentially licenses the kind of authoritarian ethics that David Harvey criticised back in 1974: an ethics of dictating to people what they can and cannot do by appeal to supposed 'natural imperatives'. Thus, far from overcoming a human/non-human dichotomy, this kind of holism would be seen as maintaining it in order to discipline people's actions regarding the earth's environmental systems.

CONCLUSION

I argued in Chapter 1 and at the end of Chapter 4 that the 'geographical experiment' is over in all but name. The 'after-natural' thinking discussed in the previous pages can, despite this, be seen as a minority attempt to renew that experiment while superceding the vocabulary of 'society' and 'nature' that underpinned it during Mackinder's time. If taken seriously, the challenge of post-natural thinking is a profound one for geography, as well as for everyday conceptions of nature in the world outside. It involves nothing less than a questioning of the division of academic labour currently organising geography as a research and teaching discipline. If pursued to its logical conclusion it means that human geographers could no longer study 'human' phenomena alone, nor physical geographers biophysical phenomena alone. In effect, environmental geography – currently the smallest of the discipline's three main branches – would colonise the whole space of the discipline but in a way different to how it is currently practised. Geographers of all stripes would be obliged to study 'social' phenomena that are never simply social and natural phenomena that are neither asocial nor simply products of social representation and practice alone. For all sorts of reasons this geography – one attuned to a world seen as hybrid, impure, messy and mixed up – is unlikely to transpire. Advocated by a minority in the discipline, it is unlikely to alter the research and teaching practices of the majority, despite its merits. Even so, there *are* some reasons for optimism. These days, both the wider public, research-funding bodies and university students are keen to know more about issues where the interfusions of the human and the non-human are as apparent as they are important. For instance, in Western countries there is a growing interest in organic or 'slow food' as an alternative to fast food, packaged food and large-scale, chemical intensive agriculture. By no means 'natural', this more wholesome food involves complex interactions between consumers, food suppliers, food retailers, farmers, seeds, traditional farm-animal breeds and more besides in long commodity chains. In relation to this and other issues there is the realistic prospect of environmental geographers occupying more disciplinary space within academic geography than is currently the case.

EXERCISES

- What, if any, problems do you think arise from doing away with a society–nature dualism? Put yourself in the position of a professional geographer wishing to undertake some empirical research with practical and moral implications. If you are unable to separate 'social' and 'natural' things then will this compromise your attempt to describe, explain and evaluate the world?
- What advantages arise from looking at the world relationally i.e. as a seamless continuum of different entities? You may wish to distinguish ontological, descriptive, explanatory and ethical advantages.
- Attempt to think of yourself as an 'actant' within networks that stretch across space and over time. What networks are you embedded in? What other entities share the network with you? What are some of the ways that these entities' properties are internalised by you? A good place to start is by thinking of the food networks you are implicated in each time you eat.

FURTHER READING

The four relational approaches discussed in this chapter are all now being discussed and used vigorously by particular groups of geographers. Among critical human geographers actor-network theory has arguably the most widely discussed and debated. See Castree and Macmillan (2001) and the literature cited therein for a sense of the key issues and disagreements.

6

CONCLUSION
Geography's natures

'[Nature] is something upon which very many frames of reference converge.
But there is no frame of reference which is as it were "naturally given", and
which does not have to be contended for in debate'.

(Foster 1997: 10)

In Chapter 1, I asserted that the discipline of geography has no 'nature' –
no essential, coherent character – in part because of the diverse ways in
which geographers comprehend nature. I hope the subsequent chapters
have fleshed out this assertion convincingly. I have shown that geographers
adhere to no one understanding of what nature is, no one understanding
of how it works, and no one understanding of what we should do to it. I
have shown that nature often appears in geographical discourse through
collateral concepts where it figures as a real but ghostly presence. I have
shown that different geographers study different aspects of what we happen
to call 'nature' in different ways – from the human mind and body to the
non-human world. Many human geographers, I argued, have increasingly
explained so-called 'natural' phenomena in social terms, while physical
geographers remain preoccupied with the 'realities' of natural and altered
environments – leaving environmental geographers to negotiate constructivist and realist approaches to nature as best they can. Meanwhile, a

minority of other geographers have sought to transcend the society—nature dualism that arguably underpins the different approaches to nature favoured by those located on different 'sides' of the discipline. The upshot, as I've indicated in previous pages, is that the 'geographical experiment' inaugurated over a century ago has arguably come to an end. In the current period, there is no disciplinary consensus on how to bring society and nature within one explanatory framework — not least because authors like Sarah Whatmore, Nigel Thrift and Jonathan Murdoch reject the dualistic terms in which the experiment was set up in the first place.

Geography, then, produces a diversity of knowledges about nature. It is, I would argue, an unusually wide diversity — a breadth that can be traced back to the discipline's late-nineteenth-century origins as a 'bridging subject'. In keeping with my argument in Chapter 1 that 'there is no such thing as nature', geography's nature-knowledges should be seen as part of a wider process of determining the why and wherefore of those things denoted by the concept (or one of its collateral concepts). It is important, I argued in Chapter 1, not to take these knowledges at face value. It is too easy to assume that physical (and many environmental) geographers produce 'objective' knowledge of the environment because they are scientists. Equally, it is important to ask why so many human and environmental geographers insist that what we call nature is often a social product. Finally, we need to ascertain what motivates those who argue that 'nature' is neither a social construction nor a relatively autonomous domain — which is why I devoted a section of the previous chapter to this issue. Clearly, each of geography's various research constituencies believes that their knowledge of nature tells us something important and worth knowing. But it would be naïve to defer to the claims of these constituencies simply on the grounds of their 'expertise' as groupings of highly educated university researchers. Instead, it is worth asking how that expertise is used to advance particular claims about what is (and is not) natural. That sentiment applies, incidentally, to this book. I have made my own argument about nature here and used my position as a professional academic to do so. But it would be inconsistent of me to scrutinise the views of other geographers on nature without acknowledging that my own deserve equally close examination.

This last comment bears directly on student readers of this book. If you've made it this far your understanding of how and why geographers study nature will, I hope, have been challenged. Many students opt for a geography degree because of their love of nature, their fascination with environmental issues or their concern about environmental degradation. *Nature* will, I hope, have shown you that geographers' interest in nature extends beyond the biophysical environment. More importantly, I hope it will also have led you to grasp the fact that geographers are but one of many communities worldwide struggling to define what 'nature' is and how we should behave towards those things designated by the word. In order to make themselves heard in that struggle, geographers – like other academics – rely upon their perceived expertise. We saw this most graphically in Chapter 4 in relation to physical geographers' continued preference for describing themselves as scientists (with all that this loaded term implies). Yet the struggle does not simply go on outside universities – in the realms of environmental-policy formation, for example. It also goes on *inside* higher education too, within and between academic disciplines. Teaching is a key element of this. Whether or not they realise it, the knowledges of nature that geography students internalise while taking their degrees are part of the wider process where societal understandings of nature are shaped and moulded. Though it may appear that these knowledges are uncontestable – because they are presented to you by your professors – I've been arguing that you should see them otherwise. Education, I would argue, is politics by other means. A failure to recognise this locks students into a 'master–pupil' model of pedagogy that shackles their critical faculties.

In a book called *Teaching to Transgress*, the cultural critic Gloria Watkins (otherwise known as bell hooks) argues that both partners in the education process frequently forget what is at stake in their encounter (be it in the lecture theatre, the seminar room or, as in the present case, in the pages of a book). Misconstruing education as the simple transmission of information from one party (teachers) to another (students), these partners can fail to see the true importance of pedagogy. For Watkins, education is always life-changing for students – whether they realise it or not. There's a well-known saying that goes like this: 'as the twig is bent so the branch grows'. Along with a few other key things – such as the family and television – the education system has a major role to play in bending the twig that is a child, and in shaping the growing branch that is a teenager and a young adult. After all, by the age of twenty-one or twenty-two (the typical age of graduation from a first degree) most students in Western countries have spent some 80 per cent of their lives in full-time education. During this time, the knowledge that students assimilate is not simply 'added on' to fully formed characters – like icing on a cake or an extension to a house. Rather, that knowledge helps to mould students into certain kinds of people. Formal education cannot, in short, fail to shape the character of those who experience it.

A sober recognition of this inescapable fact is, in my view, liberating for both teachers and students at all levels of the educational system. It means, in theory at least if not necessarily in practice, that the what, the how and the why of teaching is always up for grabs. There is no one 'correct' set of things that students should know; there is no one 'proper' way of learning; there are no 'self-evident' goals of education. Instead, there are only ever *choices* about what to teach, how to teach and to what ends. This said, when these choices are made and accepted by a sufficient number of teachers then they tend to become 'common sense'. In reality, then, the content, the manner and the aims of teaching tend to become 'fixed' for long periods of time in societies like our own. Watkins's book is an attempt to remind teachers (and their students) that things could be otherwise: that together we have an 'awesome responsibility' (hooks 1994: 206) to reflect critically and frequently on what university (and pre-university) teaching is about.

I hope *Nature* has given student readers the tools to recognise that knowledges of nature are constructed and contestable. I hope they now recognise – if they did not already – that their professors (myself included) are not to be deferred to because they follow the royal road to truthful knowledge. My take on nature has been anything but neutral, even though I have seemingly 'stood back' and presented the spectrum of geographical understandings of the topic. For instance, non-representational theorists would reject my approach altogether because it focuses on knowledge as a re-presentational 'layer' that interposes itself between ourselves and the socio-natural world. The diversity of nature-knowledges within geography is enough to show that there is no one 'correct way' of understanding nature. Yet understand it we must. As the seven vignettes with which I started this book show, the topic of nature infuses our lives. This is why knowledges of nature are so important. The power to say 'this is what nature is', 'this is how it works' or 'this is how we should behave towards it' is an awesome power. The study of nature is too important to be left to geographers alone. But, as geographers, we are better equipped to undertake such study if we recognise that knowledges of nature are part of a never-ending struggle to characterise and influence the phenomena depicted in those knowledges. When it comes to nature-knowledges, the questions we must always ask of ourselves and others are these: How are these knowledges legitimated by their advocates? What sorts of realities do they seek to engender? Why do they depict nature in the ways they do? Careful answers to these questions can give us the tools to make truly informed decisions about what nature 'really is', how it functions, how to manage it and what to do with it both now and in the future.

FURTHER READING

An accessible discussion of the politics of geographical education can be found in Castree (2005b) which includes a useful further reading section.

ESSAYS AND EXAM QUESTIONS

The essay questions below can be used in examinations or as term-paper assignments. This book, along with the recommended readings at the end of each chapter, will equip students to answer most of these questions in depth with suitable direction from their course tutors. In some cases tutors may need to recommend supplementary readings. In other cases the questions below can be modified and adapted to suit a tutor's topical teaching preferences. I have not grouped the questions thematically because most of them are quite open-ended, giving students the maximum opportunity to tailor the content of their answers.

'The idea of nature is a weapon of mass distraction'. Discuss.

Do we need nature?

'What counts as nature cannot pre-exist its construction' (Braun 2002: 17). Explore the implications of this contention.

'Nature is dead! Long live nature!' Discuss.

Is nature a necessary illusion?

'Much of the moral authority that has made environmentalism so compelling as a popular movement flows from its appeal to nature as a stable external source of non-human values against which human actions can be judged without much ambiguity' (Cronon 1996: 26). Is Cronon's assessment a fair one?

'There is no such thing as nature'. Critically assess this statement.

'Whoever utters the word "nature" deserves to be needled by the question: "which nature?" ' (Beck 1995: 342). Explain and evaluate Beck's assertion.

'Nature is a chaotic concept'. To what extent do you agree with statement?

'Sticks and stones may break my bones but words will never hurt me'. Assess the applicability of this schoolyard rhyme to the word 'nature'.

'Social constructionism has helped destabilise the longstanding notion that bodies are "simply natural" or biological' (Longhurst 2000: 23). Critically evaluate social constructionist approaches to the human body.

'A person's sexual identity is given not so much by their genital anatomy as by their sexual preferences' (Wade 2002: 42). Do you agree?

'Geography is a divided discipline because human and physical geographers have entirely different understandings of nature'. Discuss.

'The one thing that is not natural is "nature"' (Soper 1995: 7). To what extent is this assertion a defensible one?

What is the relation between the nature of geography as a discipline and the nature that geographers study?

'If you're not a realist when it comes to nature then you must be a relativist'. Evaluate this claim.

'Those who embrace [the] tenets of realism will often draw arrows from the quiver of constructionism' (Gergen 2001: 16). Explain this statement.

Imagine that you are a physical geographer by profession. How would you defend the reliability of the environmental knowledge you produce if questioned by a human geographer with a 'nature-sceptical' attitude?

'Like all . . . powerful ideas, the idea of nature as wilderness – as something separate, pristine, eternal, and harmonious – has in many ways become more important than the reality it purports to describe' (Budiansky 1996: 21). Assess this contention.

'The naturalness of nature is, in one sense, inherently self-evident' (Adams 1996: 82). Is this true?

Explain some of the problems of the idea of biological essentialism using examples from *either* the human *or* the animal world.

Is nature or nurture the most important factor in explaining *either* sexual preference *or* obesity?

'Naming something gives it a reality; a name literally gives meaning to an object' (Unwin 1996: 20). Discuss this claim in relation to *either* race *or* gender.

'Meanings can mould physical responses but they are constrained by them too' (Eagleton 2000: 87). Explore how far ideas of nature can give rise to the material realities they purport merely to describe.

Compare and contrast the principal ontologies that physical geographers have employed to make sense of the natural environment.

'To dictate definition is to wield . . . power' (Livingstone 1992: 312). Explain and illustrate this contention in relation to definitions of nature.

'They cannot represent themselves; they must be represented' (Marx 1852). Explore the implications of this statement in relation to those things conventionally called 'natural'.

'Nature knows best'. Discuss.

'Concepts can only be understood in the social and intellectual circumstances in which they are employed' (Agnew et al. 1996: 10). Discuss in relation to the idea of genes.

What are the implications of SSK for our understanding of the environmental knowledge that physical geographers produce?

'The natural world does not organise itself into parables' (Cronon 1996: 50). Do you concur?

'Biology is destiny'. Assess this assertion from the perspective of *either* a feminist geographer *or* an anti-racist geographer.

'Certain biological differences exist among humans and are themselves in reality socially meaningless' (Wade 2002: 43). Using examples, explain how certain biological differences among people become socially *significant*.

'When someone . . . has the power to . . . say "this is what culture *is*" . . . and to make that meaning stick . . . then culture, as an incredibly powerful idea, is made real, as real as any other exercise of power' (Mitchell 2000: 76). Would Mitchell's statement be valid if you substituted the word 'nature' for 'culture'?

'The value of nature relies on the fact that it is *not* human' (Adams 1996: 101). Discuss.

Imagine you were *one* of the following: a deep ecologist; a researcher on the Human Genome Project; a farmer growing GM foods. How would you evaluate the following proposition: 'Nature knows best'?

How natural are 'natural hazards'?

How realistic is the environmental knowledge produced by physical geographers?

Explain the ways in which the concept of nature has functioned as *one* of the following: an ideology; a hegemonic idea; as part of a discourse. In your answer discuss the relevant theoretical framework/s and illustrate your arguments.

Can one base an ethical code on the idea of 'natural needs'?

Do physical geographers 'cut nature at the joints'? Or do they, rather, contrive their subject matter?

On what grounds can we trust the environmental knowledge produced by physical geographers?

Offer a critical analysis of the way nature is framed morally in one of the following films: *Jurassic Park, The Hulk, Gattaca, The Island of Dr. Moreau, Frankenstein, Gorillas in the Mist, Planet of the Apes.*

What can geographers learn from the 'Sokal affair'?

'In so completely denaturalising nature . . . the agency of nature . . . was denied' (Wolch and Emel 1998: xv). How applicable is this claim to recent human geography research on nature?

Write a critical review essay about one of the following books: *The Bell Curve, Taboo: Why Black Athletes Dominate Sports, A Natural History of Rape*.

'Geography has no "nature" because the nature that geographers study extends well beyond the non-human world'. Discuss.

'To explain the history of geographers' changing understandings of nature one has to look outside the discipline'. Evaluate this contention.

Human and physical geography: trial separation or permanent divorce?

'The concept of nature is politics by other means'. Using examples, discuss this assertion.

Offer a critical analysis of *one* of the following ideas: wilderness, race, the rural.

'Landforms may have epistemological value in that they facilitate the production of knowledge about the earth's physical landscape, but the question remains whether landforms have ontological status i.e. whether they consist of something more than assemblages of physical, chemical and biological properties and the relations among these properties. If, for example, it could be convincingly demonstrated that landforms are artificial constructs devised solely for method-ological convenience and that in fact the surface of the earth is a morphological con-tinuum governed by seamless spatial variations in chemical, physical and biological properties, the need for a distinct science focusing on "landforms" would be challenged' (Rhoads, 1999: 766). Do physical geographers study 'natural kinds'?

Does the 'nature of environment' pose a problem or an opportunity for a more unified physical geography?

Assess whether 'nature' or 'nurture' is the more important factor in explaining *one* of the following: physical disability, sexuality, mental illness.

'Human geographers have . . . adopted [a nature-society] dualism by seeing the biophysical environment as an irrelevant domain for addressing social issues and by advancing explanations that invoke human processes only' (Urban and Rhoads 2003: 224). Evaluate this claim and assess its implications for geography as a whole.

'In the West, we are used to thinking of the normal and the natural as one and the same' (Holloway and Hassard 2001: 5). Evaluate this contention, using examples, with reference to *either* 'human nature' *or* the non-human world.

'It is too easy for us to forget that humankind is a part of nature . . . not apart from it' (The Prince of Wales, 2002). Is Prince Charles right and does it matter?

Using real or hypothetical examples, offer a critical evaluation of actor-network theory.

Can human and physical geographers unite around new 'after-' or 'post-natural' understandings of the material world?

How does new research by 'animal geographers' challenge conventional understandings of 'nature' within and beyond geography?

NOTES

PREFACE

1 This said, there's no doubt that geographers have had little influence on other academics who study human and non-human nature. For instance, a new 'key thinkers on the environment' book lists not a single geographer among its fifty entries (Palmer, 2001). This may reflect the unoriginality of geographers' thinking about nature! But it may also reflect a hard-to-alter prejudice among non-geographers that geography is a purely empirical discipline that produces only descriptive or classificatory knowledge.

1 STRANGE NATURES

1 *Source*: 'IVF Mix-Up and the Wrong Dad', the *Guardian*, 23 August 2003.
2 *Source*: 'A Bleak Corner of Essex is Being Hailed as England's Rainforest', the *Guardian*, 3 May 2003.
3 *Sources*: R. Thornhill and C. Palmer (2000) *A Natural History of Rape* (Cambridge, Mass: MIT Press); <http://www.kenanmalik.com/reviews>
4 *Sources*: 'Cloned Foal Romps into Record Books', the *Guardian*, August 2003; 'How Noah Could Clone a New Ark', the *Observer*, 7 January 2001.
5 *Sources*: 'Fish Don't Scream', the *Guardian*, 31 July 2001; Stephen Wise (2000) *Rattling the Cage* (New York: Profile Books).
6 *Sources*: 'Southern Ocean Hunt for Ship', the *Guardian*, 19 August 2003; 'Kazakh Dam Condemns Aral Sea', the *Guardian*, 29 October 2003; 'Ice Retreats to Open North-West Passage', the *Guardian*, 11 September 2000; 'Vanishing Herbal Remedies in Need of Cure', the *Guardian*, 14 August 2001; Bjorn Lomborg (2001) *The Skeptical Environmentalist* (Cambridge: Cambridge University Press); <http://www.lomborg.com>
7 *Sources*: 'Revealed: The Secret of Human Behaviour', the *Observer*, 11 February 2001; 'The Science Behind Racism', the *Guardian*, 10 May 2000; John Entine (2000) *Taboo* (Washington, DC: Public Affairs Publications).

8 I will say more about science and the study of nature in Chapter 4 – specifically with regard to physical geography.

9 I shall discuss Braun's research in more detail in Chapter 3.

2 THE 'NATURE' OF GEOGRAPHY

1 This kind of post-Davis physical geography was redolent of the research of one of Davis's contemporaries: G.K. Gilbert. Gilbert is often held up as a pioneer of evidence-based, 'scientific' physical geography focusing on smaller spatial and temporal scales of analysis.

2 I enter this qualifier because many doubt whether geographers of this period ever signed up to a common method as opposed to a looser understanding of the 'proper' way to interrogate the real world.

3 I will say much more about scientific method in Chapter 4.

4 Cultural ecology was also a formative influence on Third World political ecology, especially in North America – see Robbins (2004a).

5 When I use the word 'social' in social construction it is in a generic way to refer to economic, cultural and political processes that impinge upon those things we call 'nature'. As will be seen in Chapter 3, the precise meaning of 'social' in social construction varies from geographer to geographer depending on their theoretical perspective and the empirical focus of their research.

6 Oxford University Press have a series of easy-to-read 'Very Short Introductions' that cover the three 'posts' by Chris Butler, Catherine Belsey and Robert Young respectively.

7 Note that human geography's 'cultural left' and 'social left' are not synonymous with the subfield of social and cultural geography. Rather, they cross-cut virtually all of the subdisciplines comprising human geography.

8 I use the word 'realist' here not just in the specialist sense of 'transcendental realism' discussed earlier in the chapter but in the wider sense of a belief in the existence of a non-human world that is different from and irreducible to societal representations and manipulations.

3 DE-NATURALISATION: BRINGING NATURE BACK IN

1 I should confess that my previous writings on the topic of nature would probably be categorised as 'de-naturalising' in character. Indeed, later in this chapter I cite one of my published essays as part of a discussion of the material (or physical) construction of the non-human world.

2 Though Jared Diamond is one of the few notable present-day geographers that proves the rule.

3 Perhaps the most outspoken geographical defender of scientific truth over falsity is the biogeographer Phillip Stott. His personal website punctures what he sees as the falsehoods perpetrated by environmentalists. See <http://www.probiotech.fsnet.co.uk>.

4 Few in geography have used Foucault's ideas to make sense of how the non-human world is discursively constructed. An edited book by Darier (1999) examines environmental discourses, though no geographers contributed to this volume. In his more recent work Braun (2000) draws upon Foucault more explicitly, while Demeritt makes use of his ideas in an analysis of how forests are subject to 'scientific management' (Demeritt 2001a).

5 In the main, research by critical human geographers into material constructions of nature has drawn upon political economy for theoretical inspiration. The term 'political economy' describes a cluster of theories which offer a critical understanding of how economies work, focusing on power and the unequal distribution of wealth among other things (see Caparaso and Levine 1992). Marxism is a political-economic theory of prime importance within and beyond human and environmental geography. I mention this because critical human geographers have largely ignored *social theory* in their investigations of nature. The term social theory describes a cluster of approaches that analysis the constitution of societies in terms of their characteristic social relations, principal social groups, and main forms of power and resistance. Though political economy and social theory overlap, they are not synonymous (see Goldblatt 1996). Outside geography, critical researchers have adapted social theory to questions of the environment (especially in sociology) – quite why critical geographers have ignored these researchers' work is unclear.

Though I examine material constructionism in this section of the chapter, a small number of Marxists in geography have analysed the construction of nature at the level of *both* representation and materiality. This is not to say that representations aren't material (this, after all, is my main argument in this book). What I mean is that some Marxist geographers have sought to link representations of nature to an understanding of how the 'real natures' they refer to are transformed in the interests of certain economic classes. For instance, using the a 'regulation theory' framework, Gavin Bridge has shown how businesses and state institutions put a very particular 'spin' on the way the natural environment is utilised in capitalist societies (see Bridge 1998; Bridge and McManus 2000). Meanwhile, George Henderson (1999) has used a specific conception of 'ideology' (see Box 3.1) to analyse how the transformation of the Californian environment in the early twentieth century was refracted through novels, pamphlets and other written media of the time.

4 TWO NATURES? THE DIS/UNITY OF GEOGRAPHY

1 Unless otherwise specified I am thus not using the term 'realism' in the highly specific sense meant by transcendental or critical realists (as discussed in Chapter 2).

2 In this chapter's discussion of physical geography I do not want to create the false impression that the field is somehow coherent or unified in the way it investigates and understands the non-human world. Contemporary physical geography is a diverse, some would say fractured, field of research and

teaching (Gregory et al. 2002). Because of space constraints I cannot convey this diversity and division here.

3 This is not to imply that physical geographers never undertake laboratory experiments or laboratory analysis of data gathered in the field. They frequently do both, but the ultimate aim is to use laboratory study in order to understand the real environment.

4 Though this niche is an important one, physical geographers often feel that other natural scientists are suspicious about the rigour of their research. There is a perception, some physical geographers maintain, that their field is deemed a 'lesser science' than, say, physics or chemistry.

5 For instance, Thornes and McGregor (2003) make this argument in relation to the study of climate.

6 Though *not* physical geographers and physical geography thus far, excepting Demeritt's work.

7 Scientists also frequently ask 'what?' questions, relating less to causes and effects (i.e. explanation) and more to *describing* the nature of a phenomenon that is little known.

8 Haines-Young and Petch (1986: ch. 1) define the terms 'model', 'theory' and 'law' in the physical-geography context with greater precision than I can here.

9 Rather as Forsyth did in his research in the Himalayas – refer back to the section 'Re-presenting nature' in Chapter 3.

10 The term 'retroduction' is also used to describe a process of identifying past events and processes on the basis of present-day evidence. So-called 'abduction' is also an important aspect of many physical geographers' investigations. This is discussed in Box 4.6.

11 Feyerabend is often seen as one of the inspirations for SSK.

12 These four bodies of thought about how the non-human world operates, while being far from identical, emphasise the non-linearity, unpredictability and irregularity of biophysical phenomena at a number of spatio-temporal scales.

13 There are interesting debates in physical geography as to whether the field is – or should be – value-free (see, for example, the *Annals of the Association of American Geographers*, 1998). In addition, a few physical geographers have called for the formal inclusion of ethical arguments in their research – see, for example, Richards (2003b).

BIBLIOGRAPHY

Abramovitz, J. (2001) 'Averting unnatural disasters' in L. Brown (ed.) *State of the World* (London: Earthscan) pp. 123–43.

Ackerman, E. (1945) 'Geographic training, wartime research and immediate professional interests', *Annals of the Association of American Geographers* 35: 121–43.

Adams, W.M. (1996) *Future Nature: A Vision for Conservation* (London: Earthscan).

Adams, W.M. and M. Mulligan (2002) (eds) *Decolonising Nature: Strategies for Conservation in a Postcolonial Era* (London: Earthscan).

Agnew, J., D. Livingstone and A. Rogers (eds) (1996) *Human Geography* (Oxford: Blackwell).

Anderson, K. (2001) 'The nature of "race"' in N. Castree and B. Braun (eds) *Social Nature* (Oxford: Blackwell) pp. 64–83.

Annals of the Association of American Geographers (1998) 'Science, policy and ethics forum', 88, 2: 277–310.

Antipode (2005) 'Symposium on *Hybrid Geographies*', 37, 5.

Bagnold, R.A. (1941) *The Physics of Blown Sand and Desert Dunes* (London: Methuen).

Baker, V. (1999) 'The pragmatic roots of American Quaternary geology and geomorphology', *Geomorphology* 16, 2: 197–215.

Bakker, K. and G. Bridge (2003) 'Material worlds? Revisiting the matter of nature', unpublished paper, available from the authors.

Banton, M. (1998) *Racial Theories* (Cambridge: Cambridge University Press).

Barker, C. (2000) *Cultural Studies* (London: Sage).

Barrett, L. (1999) 'Particulars in context', *Annals of the Association of American Geographers* 89, 4: 707–12.

Barrett, M. (1992) *The Politics of Truth* (Cambridge: Polity).

Barrows, H. (1923) 'Geography as human ecology', *Annals of the Association of American Geographers* 13, 1: 1–4.

Barry, J. (1999) *Environment and Social Theory* (London: Routledge).

Bartram, R. and S. Shobrook (2000) 'Endless/end-less natures', *Annals of the Association of American Geographers* 90, 2: 370–80.

Bassett, K. (1999) 'Is there progress in human geography?', *Progress in Human Geography* 23, 1: 27–47.

Bassett, T. and K.B. Zueli (2000) 'Environmental discourses and the Ivorian savanna', *Annals of the Association of American Geographers* 90, 1: 67–95.

Bateman, I. and R.K. Turner (1994) *Environmental Economics: An Elementary Introduction* (London: Harvester Wheatsheaf).

Battarbee, R., R. Flower, J. Stevenson and B. Rippey (1985) 'Lake acidification in Galloway', *Nature* 314: 350–2.

Baudrillard, J. (1995) *The Gulf War Did Not Take Place* (Sydney: Power Books).

Bauer, B. (1999) 'On methodology in physical geography: a forum', *Annals of the Association of American Geographers* 89, 4: 677–778.

Beaumont, P. and C. Philo (2004) 'Environmentalism and geography: the great debate?' in J. Matthews and D. Herbert (eds) *Unifying Geography* (London: Routledge) pp. 94–116.

Beck, U. (1992) *Risk Society* (London: Sage).

Bennett, J. and W. Chaloupka (eds) (1993) *In the Nature of Things: Language, Politics, and the Environment* (Minneapolis, Minn.: University of Minnesota Press).

Bennett, R. and R. Chorley (1978) *Environmental Systems: Philosophy, Analysis and Control* (London: Methuen).

Benton, T. (1994) 'Biology and social theory in the environmental debate' in T. Benton and M. Redclift (eds) *Social Theory and the Global Environment* (London: Routledge) pp. 28–50

Bird, E.A. (1987) 'Social constructions of nature', *Environmental Review* 11, 2: 255–64.

Bird, J. (1989) *The Changing Worlds of Geography* (Oxford: Oxford University Press).

Blaikie, P. (1985) *The Political Economy of Soil Erosion in Developing Countries* (London: Longman).

Blaikie, P. and H. Brookfield (1987) *Land Degradation and Society* (London: Longman).

Blaikie, P., T. Cannon, I. Davis and B. Wisner (1994) *At Risk: Natural Hazards, People's Vulnerability, and Disasters* (London: Routledge).

Blunt, A. and J. Wills (2000) *Dissident Geographies* (Harlow: Pearson).

Botkin, D. (1990) *Discordant Harmonies* (Oxford: Oxford University Press).

Boyd, W., W.S. Prudham, and R. Schurman (2001) 'Industrial dynamics and the problem of nature', *Society and Natural Resources* 14, 4: 555–70.

Bradbury, I., J. Boyle and A. Morse (2002) *Scientific Principles for Physical Geographers* (Harlow: Prentice Hall).

Bradley, P.N. (1986) in R.J. Johnston and P. Taylor (eds) (1986) *A World in Crisis?* (Oxford: Blackwell) pp. 89–106.

Braithewaite, R.B. (1953) *Scientific Explanation* (Cambridge: Cambridge University Press).

Braun, B. (2000) 'Producing vertical territory', *Ecumene* 7, 1: 7–46.

Braun, B. (2002) *The Intemperate Rainforest* (Minneapolis, Minn: Minnesota University Press).

Braun, B. (2004) 'Nature and culture: on the career of a false problem' in J. Duncan et al. (eds) *A Companion to Cultural Geography* (Oxford: Blackwell) pp. 150–79.

Braun, B. and J. Wainwright (2001) 'Nature, poststructuralism and politics' in N. Castree and B. Braun (eds) *Social Nature* (Oxford: Blackwell) pp. 41–63.

Bridge, G. (1998) 'Excavating nature: environmental narratives and discursive regulation in the mining industry', in A. Herod, G. O'Tuathail and S. Roberts (eds) *An Unruly World?* (New York: Routledge) pp. 219–44.

Bridge, G. (2000) 'The social regulation of resource access and environmental impact: cases from the U.S. copper industry', *Geoforum* 31, 2: 237–56.

Bridge, G. and P. McManus (2000) 'Sticks and stones: environmental narratives and discursive regulation in the forestry and mining sectors', *Antipode* 32, 1: 10–47.

Brunsden, D. and J. Thornes (1979) 'Landscape sensitivity and change', *Transactions of the Institute of British Geographers* 4, 4: 463–84.

Budiansky, S. (1996) *Nature's Keepers* (London: Orion Books).

Bunce, M. (2003) 'Reproducing rural idylls', in P. Cloke (ed.) *Country Visions* (Harlow: Prentice Hall) pp. 14–28.

Bunge, W. (1962) *Theoretical Geography* (Lund: C. Gleerup).

Burgess, J. (1992) 'The cultural politics of nature conservation and economic development' in K. Anderson and F. Gale (eds) *Inventing Places* (Sydney: Longman) pp. 235–52.

Burgess, J., C. Harrison and P. Filius (1998) 'Environmental communication and the cultural politics of environmental citizenship', *Environment and Planning A* 30, 10: 1445–60.

Burley, J.C. and J. Harris (eds) (2002) *A Companion to Genetics: Philosophy and the Genetic Revolution* (Oxford: Basil Blackwell).

Burningham K. and G. Cooper (1999) 'Being constructive: social constructionism and the environment', *Sociology* 33, 2: 297–316

Burr, V. (1995) *An Introduction to Social Constructionism* (London: Routledge).

Burt, T.P. (2003a) 'Realms of gold, wild surmise and wondering about physical geography' in S. Trudgill and A. Roy (eds) *Contemporary Meanings in Physical Geography* (London: Arnold) pp. 49–61.

Burt, T.P. (2003b) 'Upscaling and downscaling in physical geography' in S. Holloway et al. (eds) *Key Concepts in Geography* (London: Sage) pp. 209–27.

Burt, T.P. (2005) 'General/particular', in N. Castree, A. Rogers and D. Sherman (eds) *Questioning Geography* (Oxford: Blackwell) pp. 185–200.

Burt, T.P. and D. Walling (eds) (1984) *Catchment Experiments in Fluvial Geomorphology* (Norwich: Geobooks).

Burt, T.P., L. Heathwaite and S. Trudgill (eds) (1993) *Nitrates: Processes, Patterns and Control* (Chichester: Wiley).

Burton, I. (1963) 'The quantitative revolution and theoretical geography', *The Canadian Geographer* 7: 151–62.

Butler, R. and H. Parr (eds) (1999) *Mind and Body Spaces* (London: Routledge).

Callicott, J.B. and M. Nelson (eds) (1998) *The Great New Wilderness Debate* (Athens, Ga.: University of Georgia Press).

Caparaso, J. and D. Levine (1992) *Theories of Political Economy* (Cambridge: Cambridge University Press).

Carson, R. (1962) *Silent Spring* (London: Hamilton).

Castree, N. (1997) 'Nature, economy and the cultural politics of theory: the "war against the seals" in the Bering Sea, 1870–1911', *Geoforum* 28, 1: 1–20.

Castree, N. (2000) 'The production of nature', in E. Sheppard and T. Barnes (eds) *A Companion to Economic Geography* (Oxford: Blackwell) pp. 269–75.

Castree, N. (2001a) 'Marxism, capitalism and the production of nature' in N. Castree and B. Braun (eds) *Social Nature* (Oxford: Blackwell) pp. 189–207.

Castree, N. (2001b) 'Socializing nature' in N. Castree and B. Braun (eds) *Social Nature* (Oxford: Blackwell) pp. 1–21.

Castree, N. (2005a) 'Is geography a science?' in N. Castree, A. Rogers and D. Sherman (eds) *Questioning Geography* (Oxford: Blackwell) pp. 110–25.

Castree, N. (2005b) 'Whose geography?', in N. Castree, A. Rogers and D. Sherman (eds) *Questioning Geography* (Oxford: Blackwell) pp. 240–52.

Castree, N. and B. Braun (2001) *Social Nature* (Oxford: Blackwell).

Castree, N. and T. Macmillan (2001) 'Dissolving dualisms' in N. Castree and B. Braun (eds) *Social Nature* (Oxford: Blackwell) pp. 208–25.

Castree, N. and B. Braun (2004) 'Constructing rural natures' in P. Cloke et al. (eds) *Handbook of Rural Studies* (London: Sage) pp. 161–72.

Chalmers, A. (1999) *What is This Thing Called Science?* 3rd edn (Buckingham: Open University Press).

Chamberlin, T.C. (1965) [1980] 'The method of multiple working hypotheses', *Science* 148: 754–9.

Chisholm, G. (1889) *Handbook of Commercial Geography* (London: Longman, Green & Co.).

Chorley, R. (ed.) (1969) *Water, Earth and Man* (London: Methuen).

Chorley, R. and B. Kennedy (1971) *Physical Geography: A Systems Approach* (London: Prentice Hall).

Church, M. (1996) 'Space, time and the mountain: how do we order what we see?' in B. Rhoads and C. Thorne (eds) *The Scientific Nature of Geomorphology* (Chichester: Wiley) pp. 147–70.

Clements, F. (1916) *Plant Succession* (Washington, DC: Carnegie Institution).

Clifford, N. (2001) 'Physical geography – the naughty world revisited', *Transactions of the Institute of British Geographers* 26, 4: 387–9.

Cloke, P., C. Philo and D. Sadler (1991) *Approaching Human Geography* (London: Paul Chapman).

Collingwood, R.G. (1945) *The Idea of Nature* (Oxford: Clarendon Press).

Collins, H. (1985) *Changing Order* (London: Sage).

Conley, V.A. (1997) *Ecopolitics: The Environment in Poststructuralist Thought* (London: Routledge).

Cooke, R. (1992) 'Common ground, shared inheritance', *Transactions of the Institute of British Geographers* 17, 2: 131–51.

Cosgrove, D. (1984) *Social Formation and Symbolic Landscape* (London: Croom Helm).

Cosgrove, D. and S. Daniels (1988) *Iconography of Landscape* (Cambridge: Cambridge University Press).

Cosgrove, D. and P. Jackson (1987) 'New directions in cultural geography', *Area* 19, 2: 95–101.

Cox, K. and R. Golledge (eds) (1969) *Behavioural Problems in Geography* (Evanston, Ill.: Northwestern University).

Cronon, W. (1996) 'In search of nature' in W. Cronon (ed.) *Uncommon Ground* (New York: W.W. Norton) pp. 23–68.

Cronon, W. (1996b) 'The trouble with wilderness; or, getting back to the wrong nature', in W. Cronon (ed.) *Uncommon Ground* (New York: W.W. Norton) pp. 69–90.

Cumberland, K. (1947) *Soil Erosion in New Zealand* (Wellington: Whitcomb & Tombs).

Darier, E. (ed.) (1999) *Discourses of the Environment* (Oxford: Blackwell).

Davidson, D. (1978) *Science for Physical Geographers* (London: Arnold).

Davis, W.M. (1906) 'An inductive study of the content of geography', *Bulletin of the American Geographical Society* 38: 67–84.

Demeritt, D. (1996) 'Social theory and the reconstruction of science and geography' *Transactions of the Institute of British Geographers* 21, 4: 484–503.

Demeritt, D. (1998) 'Science, social constructivism, and nature', in B. Braun and N. Castree (eds) *Remaking Reality: Nature at the Millennium* (London: Routledge) pp. 177–97.

Demeritt, D. (2001a) 'Scientific forest conservation and the statistical picturing of nature's limits', *Society and Space* 19, 4: 431–60.

Demeritt, D. (2001b) 'Being constructive about nature', in N. Castree and B. Braun (eds) *Social Nature* (Oxford: Blackwell) pp. 22–40.

Demeritt, D. (2001c) 'The construction of global warming and the politics of science', *Annals of the Association of American Geographers* 91, 3: 307–37.

Demeritt, D. (2001d) 'Science and the understanding of science', *Annals of the Association of American Geographers* 91, 2: 345–9.

Demeritt, D. (2002) 'What is "the social construction of nature?"', *Progress in Human Geography* 26, 6: 767–90.

Dickens, P. (2000) *Social Darwinism* (Buckingham: Open University Press).

Duncan, J., N. Johnson and R. Schein (eds) (2004) *A Companion to Cultural Geography* (Oxford: Blackwell).

Duncan, N. (ed.) (1996) *BodySpace: Destabilizing Geographies of Gender and Sexuality* (London: Routledge).

Duster, T. (1990) *Backdoor to Eugenics* (London: Routledge).

Eagleton, T. (2000) *The Idea of Culture* (Oxford: Blackwell).

Earle, C., K. Mathewson and M. Kenzer (eds) (1996) *Concepts in Human Geography* (Lanham: Rowman & Littlefield).

Eckholm, E. (1976) *Losing Ground* (New York: W.W. Norton).

Eden, S. (1996) 'Public participation in environmental policy', *Public Understanding of Science* 5, 2: 183–204.

Eden, S. (2003) 'People and the contemporary environment', in R. Johnston and M. Williams (eds) *A Century of British Geography* (Oxford: Oxford University Press) pp. 213–43.

Edgar, A. and P. Sedgwick (2002) *Cultural Theory: The Key Thinkers* (London: Routledge).

Ehrlich, P. (1970) *The Population Bomb* (New York: Ballantine Books).

Emel, J. and R. Peet (1989) 'Resource management and natural hazards' in R. Peet and N. Thrift (eds) *New Models in Geography* Vol. I (Boston, Mass.: Unwin Hyman) pp. 49–76.

Endfield, G. (2004) *Environment* (London: Routledge).

Entine, J. (2000) *Taboo: Why Black Athletes Dominate Sports and Why We're Afraid to Talk About It* (Washington, DC: Public Affairs Publications).

Entrikin, N. (1976) 'Contemporary humanism in geography', *Annals of the Association of American Geographers* 66, 4: 615–32.

Entrikin, N. (1979) 'Philosophical issues in the scientific study of regions' in D.T. Herbert and R.J. Johnston (eds) *Geography and the Urban Environment* Vol. IV (Chichester: Wiley) pp. 1–27.

Fairhead, J. and M. Leach (1996) *Misreading the African Landscape* (Cambridge: Cambridge University Press).

Favis-Mortlock, D. and D. de Boer (2003) 'Simple at heart?' in S. Trudgill and A. Roy (eds) *Contemporary Meanings in Physical Geography* (London: Arnold) pp. 127–72.

Feyeraband, P. (1975) *Against Method* (London: Verso).

Findlay, A. (1995) 'Population crises?' in R.J. Johnston et al. (eds) *Geographies of Global Change* (Oxford: Blackwell) pp. 152–74.

Fitzsimmons, M. (1989) 'The matter of nature', *Antipode* 21, 2: 106–20.

Fleure, H. (1919) 'Human regions', *Scottish Geographical Magazine* 35: 31–45.

Fleure, H. (1926) *Wales and Her People* (Wrexham: Hughes & Son).

Forsyth, T. (1996) 'Science, myth and knowledge', *Geoforum* 27, 3: 375–92.

Forsyth, T. (2003) *Critical Political Ecology* (London: Routledge).

Foster, J. (ed.) (1997) *Valuing Nature?* (London: Routledge).

Foucault, M. (1979) *The History of Sexuality* (London: Allen Lane).

Fuss, D. (1989) *Essentially Speaking: Feminism, Nature and Difference* (London and New York: Routledge).

Gardner, R. (1996) 'Developments in physical geography', in E. Rawling and R. Daugherty (eds) *Geography into the 21st Century* (Chichester: Wiley) pp. 26–38.

Gergen, K. (2001) *Social Construction in Context* (London: Sage).

Gibbs, D. (2000) 'Ecological modernisation, regional economic development and Regional Development Agencies', *Geoforum* 31, 9–19.

Gieryn, T. (1983) 'Boundary work and the demarcation of science from non-science', *American Sociological Review* 48, 5: 781–95.

Gifford, T. (1996) 'The social construction of nature', *Interdisciplinary Studies in Literature and Environment* 3, 1: 27–36.

Gilbert, E.W. (1960) 'The idea of the region', *Geography* 45, 2: 157–75.

Glacken, C. (1967) *Traces on the Rhodian Shore* (Berkeley, Calif.: University of California Press).

Goldblatt, D. (1996) *Environment and Social Theory* (Cambridge: Polity Press).

Goldsmith, E., R. Allen, M. Allaby, J. Davoll and S. Lawrence (1972) *Blueprint for Survival* (Harmondsworth: Penguin).

Goodman, D. and M. Watts (eds) (1997) *Globalising Food* (New York: Routledge).

Goodman, N. (1978) *Ways of Worldmaking* (Hassocks: Harvester Press).

Goudie, A. (1984) *The Nature of the Environment: An Advanced Physical Geography* (Oxford: Blackwell).

Gould, P. (1979) 'Geography 1957–77: the Augean period' *Annals of the Association of American Geographers* 69, 1: 139–51.

Graf, W. (1979) 'Catastrophe theory as a model for change in fluvial systems' in D.D. Rhodes and G. Williams (eds) *Adjustment of the Fluvial System* (Dubuque, Ia.: Hunt) pp. 13–32.

Graf, W. (1992) 'Science, public policy and western American rivers', *Transactions of the Institute of British Geographers* 17, 1: 5–19.

Gramsci, A. (1995) *Further Selections from the Prison Notebooks* (London: Lawrence & Wishart).

Greenblatt, S. (1991) *Marvellous Possessions* (Chicago, Ill.: Chicago University Press).

Gregory, D. (1978) *Ideology, Science and Human Geography* (London: Hutchinson).

Gregory D. (1994) 'Social theory and human geography' in D. Gregory, R. Martin and G. Smith (eds) *Human Geography* (London: Macmillan) pp. 78–112

Gregory K. (2000) *The Changing Nature of Physical Geography* (London: Arnold).

Gregory, K. (2004) 'Valuing physical geography', *Geography* 89, 1: 16–25.

Gregory, K., A. Gurnell and G. Petts (2002) 'Restructuring physical geography', *Transactions of the Institute of British Geographers* 27, 2: 136–55.

Gregory, S. and D. Walling (1973) *Drainage Basin: Process and Form* (London: Arnold).

Gregory, S. and D. Walling (eds) (1974) *Fluvial Processes in Instrumented Catchments* (London: Institute of British Geographers).

Greider, T. and L. Garkovich (1994) 'Landscapes: the social construction of nature and the environment', *Rural Sociology* 59, 1: 1–24.

Gross, P.R. and N. Levitt (1994) *Higher Superstition: The Academic Left and its Quarrels with Science* (Baltimore, Md. and London: Johns Hopkins University Press).

Grosz, E. (1992) 'Bodies-cities' in B. Colomina (ed.) *Sexuality and Space* (New York: Princeton Architectural Press) pp. 45–61.

Guha, R. (1994) 'Radical American environmentalism and wilderness preservation: a Third World critique' in H. Gruen and D. Jamieson (eds) *Reflecting on Nature* (Oxford: Oxford University Press) pp. 241–51.

Guyer, J. and P. Richards (1996) 'The invention of biodiversity', *Africa* 66, 1: 1–13.

Habgood, J. (2002) *The Concept of Nature* (London: Darton, Longman & Todd).

Hacking, I. (1996) 'The disunities of the sciences' in P. Galison and D. Stump (eds) *The Disunity of Science* (Stanford, Calif.: Stanford University Press) pp. 37–74.

Haggett, P. (1965) *Locational Analysis in Human Geography* (London: Edward Arnold).

Haines-Young, R. and J. Petch (1986) *Physical Geography: Its Nature and Methods* (London: Harper & Row).

Halfon, S. (1997) 'Overpopulating the world' in P. Taylor et al. (eds) *Changing Life* (Minneapolis, Minn.: Minnesota University Press) pp. 121–48.

Haraway, D. and D. Harvey (1995) 'Nature, politics and possibilities', *Society and Space* 13, 4: 507–27.

Hardin, G. (1974) 'The ethics of a lifeboat', *Bioscience*, 24, October, 1–18.

Harrison, S. (2001) 'On reductionism and emergence in geomorphology', *Transactions of the Institute of British Geographers* 26, 3: 327–39.

Harrison, S. and P. Dunham (1998) 'Decoherence, quantum theory and their implications for the philosophy of geomorphology', *Transactions of the Institute of British Geographers* 23, 4: 501–14.

Hartshorne, R. (1939) *The Nature of Geography* (Lancaster, Pa.: Association of American Geographers).

Harvey, D. (1969) *Explanation in Geography* (London: Arnold).

Harvey, D. (1973) *Social Justice and the City* (London: Arnold).

Harvey, D. (1974) 'Population, resources and the ideology of science', *Economic Geography* 50, 2: 256–77.

Harvey, D. (1990) 'Between space and time', *Annals of the Association of American Geographers* 80, 3: 418–34.

Harvey, D. (1996) *Justice, Nature and the Geography of Difference* (Oxford: Blackwell).

Harvey, D. and D. Haraway (1995) 'Nature, politics and possibilities', *Society and Space* 13, 4: 507–27.

Heffernan, M. (2003) 'Histories of geography' in S. Holloway, S. Rice and G. Valentine (eds) *Key Concepts in Geography* (London: Sage) pp. 3–22.

Heisenberg, W. (1958) *Physics and Philosophy* (New York: Harper & Row).

Henderson, G. (1994) 'Romancing the sand: constructions of capital and nature in arid America', *Ecumene* 1, 3: 235–55.

Henderson, G. (1999) *California and the Fictions of Capital* (Oxford: Oxford University Press).

Herbertson, A.J. (1899) [1920] *Man and His Work* (London: A. & C. Black).

Herbertson, A.J. (1905) 'The major natural regions', *Geographical Journal* 25, 4: 300–12.

Herbst, J. (1961) 'Social Darwinism and the history of American geography', *Proceedings of the American Philosophical Society* 105, 6: 538–44.

Hernstein, R.J. and C. Murray (1996) *The Bell Curve: Intelligence and Class Structure in American Life* (London and New York: Free Press).

Hess, D. (1997) *Science Studies* (New York: New York University Press).

Hewitt, K. (ed.) (1983) *Interpretations of Calamity* (Boston, Mass.: Allen & Unwin).

Hinchcliffe, S. (2001) 'Indeterminacy in-decisions', *Transactions of the Institute of British Geographers* 26, 2: 182–204.

Hinchcliffe, S. (2006) *Spaces for Nature* (London: Sage).

Holliday, R. and J. Hassard (eds) (2001) *Contested Bodies* (London: Routledge).

Hollis, G.E. (1975) 'The effect of urbanization on floods', *Water Resources Research* 11, 4: 431–4.

Hollis, G.E. (1979) *Man's Impact on the Hydrological Cycle in the UK* (Norwich: Geobooks).

Holt-Jensen, A. (1999) *Geography: History and Concepts* 3rd edn (Sage: London).

Holton, G. (1993) *Science and Anti-science* (Cambridge, Mass.: Harvard University Press).

hooks, b. (1994) *Teaching to Transgress* (London: Routledge).

Horton, R.E. (1945) 'Erosional development of streams and their drainage basins', *Bulletin of the Geological Society of America* 56: 275–370.

Hubbard, P., R. Kitchin, B. Bartley and D. Fuller (2002) *Thinking Geographically* (London: Continuum).

Hudson, R. (2001) *Producing Places* (New York: Guilford).

Huntington, E. (1924) *The Character of Races* (New York: Scribner's).

Huxley, T.H. (1877) *Physiography* (London: Macmillan).

Imrie, R. (1996) *Disability and the City* (London: Paul Chapman).

Inkpen, R. (2004) *Science, Philosophy and Physical Geography* (London: Routledge)

Jacks, G. and R. Whyte (1939) *Rape of the Earth* (London: Faber & Faber).

Jackson, P. (1987) *Race and Racism* (London: Allen & Unwin).

Jackson, P. (1989) *Maps of Meaning* (London: Unwin Hyman).

Jackson, P. (1994) 'Black male: advertising and the cultural politics of masculinity', *Gender, Place and Culture* 1, 1: 49–59.

Johnston, R.J. (1986) *Philosophy and Human Geography* (London: Edward Arnold).

Johnston, R.J. (2003) 'Geography and the social sciences tradition', in S. Holloway et al. (eds) *Key Concepts in Geography* (London: Sage) pp. 51–72.

Johnston, R.J. and J. Sidaway (2004) *Geography and Geographers* 6th edn (London: Arnold).

Johnston, R.J., D. Gregory, G. Pratt and M. Watts (eds) (2000) *The Dictionary of Human Geography* 4th edn (Oxford: Blackwell).

Jones, O. (2000) '(Un)ethical geographies of human–nonhuman relations', in C. Philo and C. Wilbert (eds) *Animal Spaces, Beastly Places* (London: Routledge) pp. 268–91.

Kennedy, B. (1979) 'It's a naughty world', *Transactions of the Institute of British Geographers* 4, 4: 550–8.

Kennedy, B. (1994) 'Requiem for a dead concept', *Annals of the Association of American Geographers* 84, 4: 702–5.

Keylock, C. (2003) 'The natural science of geomorphology?' in S. Trudgill and A. Roy (eds) *Contemporary Meanings in Physical Geography* (London: Arnold) pp. 87–101.

Kirk, R. (1999) *Relativism and Reality* (London: Routledge).

Kitchin, R. (2000) *Disability, Space and Society* (Sheffield: Geographical Association).

Kloppenburg, J. (1988) *First the Seed* (Cambridge: Cambridge University Press).

Kolodny, A. (1984) *The Land Before Her* (Chapel Hill, NC: University of North Carolina Press).

Kuhn, T. (1962) *The Structure of Scientific Revolutions* (Chicago, Ill.: Chicago University Press).

Kukla, A. (2000) *Social Constructivism and the Philosophy of Science* (London: Routledge).

Lane, S. (2001) 'Constructive comments on Massey', *Transactions of the Institute of British Geographers* 26, 2: 243–56.

Lane, S. and A. Roy (2003) 'Putting the morphology back into fluvial geomorphology' in S. Trudgill and A. Roy (eds) *Contemporary Meanings in Physical Geography* (London: Arnold) pp. 103–25.

Latour, B. (1993) *We Have Never Been Modern* (Cambridge, Mass.: Harvard University Press).

Leach, M. and R. Mearns (eds) (1996) *The Lie of the Land* (New York: Heinemann).

Lechte, J. (1994) *Fifty Key Contemporary Thinkers* (London: Routledge).

Leighly, J. (1955) 'What has happened to physical geography?', *Annals of the Association of American Geographers* 45, 3: 309–18.

Leopold, L., M. Wolman and J. Miller (1964) *Fluvial Processes in Geomorphology* (San Francisco, Calif.: Freeman).

Levins, R. and R. Lewontin (1985) *The Dialectical Biologist* (Cambridge, Mass.: Harvard University Press).

Lewis, G. (2001) 'Welfare and the social construction of race' in E. Saraga (ed.) *Embodying the Social* (London: Routledge) pp. 91–138.

Light, A. and H.R. Rolston III (eds) (2003) *Environmental Ethics: An Anthology* (Oxford: Blackwell).

Liverman, D. (1999) 'Geography and the global environment', *Annals of the Association of American Geographers* 89, 1: 107–20.

Livingstone, D. (1992) *The Geographical Tradition* (Oxford: Blackwell).

Lomborg, B. *The Skeptical Environmentalist* (Cambridge: Cambridge University Press).

Longhurst, R. (2001) *Bodies* (London: Routledge).

Low, N. (ed.) (1999) *Ethics and the Global Environment* (London: Routledge).

Low, N. and B. Gleeson (1998) *Justice, Society and Nature* (London: Routledge).

Lyotard, J.-F. (1984) *The Postmodern Condition* (Minneapolis, Minn.: University of Minnesota Press).

Lynn, W. (1998) 'Animals, ethics and geography' in J. Wolch and J. Emel (eds) *Animal Geographies* (London: Verso) pp. 280–97.

McGuigan, J. (1999) *Modernity and Postmodern Culture* (Buckingham: Open University Press).

McKibben, R. (1990) *The End of Nature* (New York: Vintage).

Mackinder, H. (1887) 'On the scope and methods of geography', *Proceedings of the Royal Geographical Society* 9: 141–60.

Mackinder, H. (1902) *Britain and the British Seas* (Oxford: Clarendon Press).

Maclaughlin, J. (1999) 'The evolution of modern demography and the debate on sustainable development', *Antipode* 31, 3: 324–43.

Macnaughten, P. and J. Urry (1998) *Contested Natures* (London: Sage).

Magnusson, W. and K. Shaw (2003) *A Political Space* (Minneapolis, Minn.: University of Minnesota Press).

Malanson, G. (1999) 'Considering complexity', *Annals of the Association of American Geographers* 89, 4: 746–53.

Malik, K. (1996) *The Meaning of Race* (London: Macmillan).

Malthus, T. (1798) *Essay on the Principle of Population* (London: J. Johnson).

Mann, S. and J. Dickinson (1978) 'Obstacles to the development of a capitalist agriculture', *Journal of Peasant Studies* 5, 4: 466–81.

Marbut, C.F. (1935) *The Great Soil Groups of the World and their Development*, trans. K. D. Glinka (Ann Arbor, Mich.: Edward Bros).

Marsh, G.P. (1864) [1965] *Man and Nature*, ed. D. Lowenthal (Cambridge, Mass.: Harvard University Press).

Marshall, J. (1985) 'Geography as a scientific enterprise' in R.J. Johnston (ed.) *The Future of Geography* (London: Methuen) pp. 113–28.

Massey, D. (1999) 'Space-time, "science" and the relationship between physical and human geography', *Transactions of the Institute of British Geographers* 24, 3: 261–77.

Mayr, E. (1972) 'The nature of the Darwinian revolution', *Science* 176: 981–9.

Meadows, D., J. Randers, W.W. Behrens (1972) *Limits to Growth* (New York: Universe Books).

Merton, R. (1942) [1973] 'Science and technology in a democratic order' in N. Storer (ed.) *The Sociology of Science* (Chicago, Ill.: University of Chicago Press) pp. 267–78.

Middleton, N. (1995) *The Global Casino* (London: Arnold).

Mikesell, M. (1974) 'Geography as the study of environment' in I. Manners and M. Mikesell (eds) *Perspectives on Environment* (Washington, DC: Association of American Geographers) pp. 66–80.

Miles, R. (1989) *Racism* (London: Routledge).

Miller, A. (1931) *Climatology* (London: Methuen).

Mitchell, B. (1979) *Geography and Resource Analysis* (London: Longman).

Mitchell, D. (1995) 'There's no such thing as culture', *Transactions of the Institute of British Geographers* 20, 1: 102–16.

Mitchell, D. (2000) *Cultural Geography: A Critical Introduction* (Oxford: Blackwell).

Mitchell, T. (1988) *Colonising Egypt* (Cambridge: Cambridge University Press).

Moeckli, J. and B. Braun (2001) 'Gendered natures' in N. Castree and B. Braun (eds) *Social Nature* (Oxford: Blackwell) pp. 112–32.

Moore, D. (1996) 'Marxism, culture and political ecology' in R. Peet and M. Watts (eds) *Liberation Ecologies* (New York: Routledge) pp. 125–47.

Murdoch, J. (1997a) 'Inhuman/nonhuman/human: actor-network theory and the prospects for a nondualistic and symmetrical perspective on nature and society', *Environment and Planning D* 15, 6: 731–56.

Murdoch, J. (1997b) 'Towards a geography of heterogeneous associations', *Progress in Human Geography* 2, 3: 321–37.

Murdoch, J. and P. Lowe (2003) 'The preservationist paradox', *Transactions of the Institute of British Geographers* 28, 3: 318–33.

Myers, N. (1979) *The Sinking Ark: A New Look at the Problem of Disappearing Species* (Oxford: Pergamon).

Nash, R. (1989) *The Rights of Nature: A History of Environmental Ethics* (Madison, Wisc.: University of Wisconsin Press).

Nash, R. (2001) *Wilderness and the American Mind* 4th edn (New Haven, Conn.: Yale University Press).

Nast, H. and S. Pile (eds) (1998) *Places Through the Body* (London: Routledge).

Neuhaus, R. (1971) *In Defense of People* (New York: Macmillan).

Neumann, R. (1995) 'Ways of seeing Africa', *Ecumene* 2, 2: 149–67.

Neumann, R. (1998) *Imposing Wilderness* (Berkeley, Calif.: University of California Press).

Nietschmann, B. (1973) *Between Land and Water* (New York: Seminar Press).

Norton, B. (2000) 'Population and consumption: environmental problems as problems of scale', *Ethics and the Environment* 5, 1: 23–45.

Norwood, V. and J. Monk (eds) (1987) *The Desert is No Lady* (New Haven, Conn.: Yale University Press).

Oelschlaeger, M. (1991) *The Idea of Wilderness* (New Haven, Conn.: Yale University Press).

Okasha, S. (2002) *Philosophy of Science* (Oxford: Oxford University Press).

Ollman, B. (1993) *Dialectical Investigations* (New York: Routledge).

Olwig, K. (1996) 'Nature: mapping the ghostly traces of a concept' in C. Earle, K. Mathewson and M. Kenzer (eds) *Concepts in Human Geography* (Lanham, Md.: Rowman & Littlefield) pp. 63–96.

O'Riordan, T. (1976) *Environmentalism* (London: Pion).

O'Riordan, T. (1989) 'The challenge for environmentalism' in R. Peet and N. Thrift (eds) *New Models in Geography* Vol. I (London: Unwin Hyman) pp. 77–104.

Palmer, J. (ed.) (2001) *Fifty Key Thinkers on the Environment* (London: Routledge).

Panelli, R. (2004) *Social Geographies* (London: Sage).

Peet, R. (1999) *Modern Geographical Thought* (Oxford: Blackwell).

Peet, R. and M. Watts (eds) (1996) *Liberation Ecologies* (New York: Routledge).

Pelling, M. (2001) 'Natural hazards?' in N. Castree and B. Braun (eds) *Social Nature* (Oxford: Blackwell) pp. 170–88.

Pelling, M. (ed.) (2003) *Natural Disaster and Development in a Globalizing World* (Routledge: London).

Penrose, J. (2003) 'When all the cowboys are Indians: The nature of race in all-Indian rodeo', *Annals of the Association of American Geographers* 93, 3: 687–705.

Pepper, D. (1984) *The Roots of Modern Environmentalism* (London: Croom Helm).

Peterson, A. (1999) 'Environmental ethics and the social construction of nature', *Environmental Ethics* 21, 3: 339–57.

Petrucci, M. (2000) 'Population: time-bomb or smoke screen?', *Environmental Values* 9, 3: 325–52.

Phillips, J.D. (1999) *Earth Surface Systems: Complexity, Order and Scale* (Blackwell: Malden).

Philo, C. (2001) 'Accumulating populations: bodies, institutions and space', *International Journal of Population Geography* 7, 4: 473–90.

Philo, C. and C. Wilbert (eds) (2000) *Animal Spaces, Beastly Places* (London: Routledge).

Pickering, K. and L. Owen (1997) *An Introduction to Global Environmental Issues* (London: Routledge).

Pile, S. (1996) *The Body and the City: Psychoanalysis, Subjectivity and Space* (London: Routledge).

Pile, S. and N. Thrift (eds) (1995) *Mapping the Subject* (London: Routledge).

Pinker, S. (2002) *The Blank Slate: Denying Human Nature in Modern Life* (Harmondsworth: Penguin).

Popper, K. (1974) *Conjectures and Refutations* 5th edn (London: Routledge & Kegan Paul).

Pratt, G. and S. Hanson (1994) 'Geography and the construction of difference', *Gender, Place and Culture* 1, 1: 5–29.

Proctor, J. (1998) 'The social construction of nature', *Annals of the Association of American Geographers* 88, 3: 351–76.

Proctor, J. (2001) 'Solid rock and shifting sands' in N. Castree and B. Braun (eds) *Social Nature* (Oxford: Blackwell) pp. 225–40.

Proctor, J. and D.M. Smith (eds) (1998) *Geography and Ethics* (London: Routledge).

Pulido, L. (1996) *Environmentalism and Economic Justice: Two Chicano Struggles in the Southwest* (Phoenix, Ariz.: University of Arizona Press).

Raper, J. and D. Livingstone (2001) 'Let's get real: spatio-temporal identity and geographic entities', *Transactions of the Institute of British Geographers* 26, 2: 237–42.

Rees, J. (1990) *Natural Resources: Allocation, Economics and Policy* 2nd edn (London: Routledge).

Relph, E. (1976) *Place and Placelessness* (London: Croom Helm).

Rhoads, B. (1999) 'Beyond pragmatism: the value of philosophical discourse for physical geography', *Annals of the Association of American Geographers* 89, 4: 760–71.

Rhoads, B. and C. Thorn (1994) 'Contemporary philosophical perspectives in physical geography', *Geographical Review* 84, 1: 90–101.

Rhoads, B. and C. Thorn (eds) (1996) *The Scientific Nature of Geomorphology* (Chichester: Wiley).

Ricciardi, A. and J. Rasmussen (1999) 'Extinction rates of North American freshwater fauna', *Conservation Biology* 13: 1220–2.

Richards, K. (1990) 'Real geomorphology', *Earth Surface Processes and Landforms* 15, 2: 195–7.

Richards, K. (2003a) 'Geography and the physical sciences tradition' in S. Holloway et al. (eds) *Key Concepts in Geography* (London: Sage) pp. 23–50.

Richards, K. (2003b) 'Ethical grounds for an integrated geography' in S. Trudgill and A. Roy (eds) *Contemporary Meanings in Physical Geography* (London: Arnold) pp. 233–58.

Robbins, P. (2004a) 'Cultural ecology' in J. Duncan et al. (eds) *A Companion to Cultural Geography* (Oxford: Blackwell) pp. 180–93.

Robbins, P. (2004b) *Political Ecology* (Oxford: Blackwell).

Roberts, R. and J. Emel (1992) 'Uneven development and the tragedy of the commons', *Economic Geography* 68, 2: 249–71.

Rogers, A. (2005) 'The "nature" of geography?' in N. Castree, A. Rogers and D. Sherman (eds) *Questioning Geography* (Oxford: Blackwell) pp. 12–25.

Rose, G. (1993) *Feminism and Geography* (Oxford: Blackwell).

Rose, G., V. Kinnaird, M. Morris and C. Nash (1997) 'Feminist geographies of environment, nature and landscape' in Women and Geography Study Group *Feminist Geographies* (Harlow: Longman) pp. 146–90.

Rose, S. and R. Lewontin (1990) *Not in Our Genes: Biology, Ideology and Human Nature* (London: Penguin).

Ross, A. (1994) *The Chicago Gangster Theory of Life* (London: Verso).

Ross, A. (ed.) (1996) *Science Wars* (Durham, NC: Duke University Press).

Rothenberg, D. (ed.) (1995) *Wild Ideas* (Minneapolis, Minn.: University of Minnesota Press).

Rowles, G. (1978) *The Prisoners of Space* (Boulder, Colo.: Westview).

Russell, R.J. (1949) 'Geographical geomorphology', *Annals of the Association of American Geographers* 39, 1: 1–11.

Saarinen, T. (1966) *Perception of the Drought Hazard on the Great Plains* (Chicago, Ill.: Chicago University Press).

Said, E. (1978) *Orientalism* (New York: Vintage).

Saraga, E. (2001) 'Abnormal, unnatural and immoral? The social construction of sexualities' in E. Saraga (ed.) *Embodying the Social* (London: Routledge) pp. 139–88.

Sardar, Z. and B. van Loon (2002) *Introducing Science* (Lanham. Md.: Totem Books).

Sauer, C. (1925) 'The morphology of landscape', *University of California Publications in Geography* 2, 19–53.

Sauer, C. (1956) 'The agency of man on earth' in W.I. Thomas (ed.) *Man's Role in Changing the Face of the Earth* (Chicago, Ill.: Chicago University Press) pp. 49–69.

Sayer, A. (1984) *Method in Social Science* (London: Hutchinson).

Sayer, A. (1993) 'Postmodernist thought in geography: a realist view', *Antipode* 25, 3: 320–44.

Schaefer, F. (1953) 'Exceptionalism in geography', *Annals of the Association of American Geographers* 43, 3: 226–49.

Scheidegger, A.E. (1961) *Theoretical Geomorphology* (Berlin: Springer-Verlag).

Schneider, S. (2001) 'A constructive deconstruction of deconstructionists', *Annals of the Association of American Geographers* 91, 2: 338–45.

Schumm, S. (1979) 'Geomorphic thresholds', *Transactions of the Institute of British Geographers* 4, 4: 485–515.

Schumm, S. (1991) *To Interpret the Earth* (Cambridge: Cambridge University Press).

Schumm, S. and R.W. Lichty (1965) 'Time, space and causality', *American Journal of Science* 263: 110–19.

Seamon, D. and R. Mugerauer (eds) (1985) *Dwelling, Place and Environment* (Dordrecht: Martinus Nijhoff).

Segal, L. (1997) 'Sexualities' in K. Woodward (ed.) *Identity and Difference* (London: Sage) pp. 183–238.

Semple, E. (1911) *Influences of Geographic Environment* (New York: Henry Holt).

Sherman, D. (1996) 'Fashion in geomorphology' in B. Rhoads and C. Thorn (eds) *The Scientific Nature of Geomorphology* (Chichester: Wiley) pp. 87–114.

Shilling, C. (1997) 'The body and difference' in K. Woodward (ed.) *Identity and Difference* (London: Sage) pp. 63–120.

Shilling, C. (2003) *The Body and Social Theory* 2nd edn (London: Sage).

Simmons, I.G. (1990) 'No rush to grown green', *Area* 22, 4: 384–7.

Simpson, G. (1963) 'Historical science' in C. Albritton (ed.) *The Fabric of Geology* (Reading, Mass.: Addison-Wesley) pp. 24–48.

Sims, P. (2003) 'Trends and fashions in physical geography' in S. Trudgill and A. Roy (eds) *Contemporary Meanings in Physical Geography* (London: Arnold) pp. 3–24.

Sismondo, S. (2003) *An Introduction to Science and Technology Studies* (Oxford: Blackwell).

Slaymaker, O. and T. Spencer (1998) *Physical Geography and Global Environmental Change* (Harlow: Longman).

Smith, J.R. (1913) *Industrial and Commercial Geography* (New York: Henry Holt & Co.).

Smith, M.J. (2002) *Social Science in Question* (London: Sage).

Smith, N. (1984) *Uneven Development* (Oxford: Blackwell).

Smith, N. (1996) 'The production of nature' in G. Robertson, M. Mash, L. Tickner, J. Bird, B. Curtis and T. Putnam (eds) *FutureNatural* (London: Routledge) pp. 35–54.

Smith, N. (1987) 'Academic war over the field of geography', *Annals of the Association of American Geographers* 77, 2: 155–72.

Smith, R. (2003) 'Baudrillard's non-representational theory', *Environment and Planning D: Society and Space* 21, 1: 67–84.

Snyder, G. (1996) 'Nature as seen from Kitkitdizze is no "social construction"', *Wild Earth* 6: 8–9.

Somerville, M. (1848) *Physical Geography* (London: Murray).

Soper, K. (1995) *What Is Nature?* (Oxford: Blackwell).

Soper, K. (1996) 'Nature/nature' in G. Robertson, M. Mash, L. Tickner, J. Bird, B. Curtis and T. Putnam (eds) *FutureNatural* (London: Routledge) pp. 22–34.

Soule, G. and G. Lease (eds) (1995) *Reinventing Nature? Responses to Postmodern Deconstruction* (Washington, DC: Island Press).

Spedding, N. (2003) 'Landscape and environment' in S. Holloway et al. (eds) *Key Concepts in Geography* (London: Sage) pp. 281–304.

Sugden, D. (1996) 'The east Antarctic ice sheet: unstable ice or unstable ideas?', *Transactions of the Institute of British Geographers* 21, 4: 443–54.

Sugden, D., M. Summerfield and T. Burt (1997) 'Linking short-term processes and landscape evolution', *Earth Surface Processes and Landforms* 22, 2: 193–4.

Stamp, D. (1957) 'Major natural regions', *Geography* 42, 3: 201–16.

Stoddart, D. (1986) *On Geography and its History* (Oxford: Blackwell).

Strahler, A.N. (1952) 'Dynamic basis of geomorphology', *Bulletin of the Geological Society of America* 63: 923–37.

Sullivan, S. (2000) 'Getting the science right, or introducing science in the first place?' in P. Stott and S. Sullivan (eds) *Political Ecology: Science, Myth and Power* (London: Arnold) pp. 15–44.

Takacs, J. (1996) *The Idea of Biodiversity* (Baltimore, Md.: Johns Hopkins University Press).

Taylor, P. and R. Garcia-Barrios (1999) 'The dynamics of socio-environmental change and the limits of neo-Malthusian environmentalism' in M. Dore and T. Mount (eds) *Global Environmental Economics* (Oxford: Blackwell) pp. 139–67.

Thomas, D. and A. Goudie (eds) (2000) *The Dictionary of Physical Geography* 3rd edn (Oxford: Blackwell).

Thomas, D.S.G. and N.J. Middleton (1994) *Desertification: Exploding the Myth* (Chichester: Wiley).

Thomas, W.L. (ed.) (1956) *Man's Role in Changing the Face of the Earth* (Chicago, Ill.: Chicago University Press).

Thompson, M., M. Warburton and M. Hatley (1986) *Uncertainty on a Himalayan Scale* (London: Milton Ash Publications).

Thorne, J. (2003) 'Change and stability in environmental systems' in S. Holloway et al. (eds) *Key Concepts in Geography* (London: Sage) pp. 131–50.

Thornes, J. and G. McGregor (2003) 'Cultural climatology' in S. Trudgill and A. Roy (eds) *Contemporary Meanings in Physical Geography* (London: Arnold) pp. 173–98.

Thornhill, R. and C. Palmer (2000) *A Natural History of Rape* (Cambridge, Mass.: MIT Press).

Thrift, N. (1996) *Spatial Formations* (London: Sage).

Thrift, N. (2003) 'Summoning life' in P. Cloke et al. (eds) *Envisioning Human Geography* (London: Arnold) pp. 65–77.

Thrift, N. and S. Pile (eds) (1995) *Mapping the Subject* (London: Routledge).

Trigg, R. (1988) *Ideas of Human Nature: An Historical Introduction* (Oxford: Basil Blackwell).

Tuan, Y.-F. (1974) *Topophilia: A Study of Environmental Perception, Attitudes and Values* (Englewood Cliffs, NJ: Prentice Hall).

Turner II, B.L., W.C. Clark, R.W. Kates, J.F. Richards, J.T. Mathews and W.B. Meyer (1990) *The Earth as Transformed by Human Action* (Cambridge: Cambridge University Press).

Turner II, B.-L., W.C. Clark, R.W. Kates, J.F. Richards, J.T. Matthews and W.B. Meyer (1990) *The Earth as Transformed by Human Action* (Cambridge: Cambridge University Press).

Turner II, B.-L. (2002) 'Contested identities: human-environment geography and disciplinary implications in a restructuring academy', *Annals of the Association of American Geographers* 92, 1: 52–74.

Twidale, C. (1983) 'Scientific research – some methodological problems', *The New Philosophy* 86, 1: 51–66.

Unwin, T. (1992) *The Place of Geography* (Harlow: Longman).

Unwin, T. (1996) 'Academic geography: key questions for discussion' in E. Rawling and R. Daugherty (eds) *Geography into the 21st Century* (Chichester: Wiley) pp. 18–25.

Urban, M. and B. Rhoads (2003) 'Conceptions of nature' in S. Trudgill and A. Roy (eds) *Contemporary Meanings in Physical Geography* (London: Arnold) pp. 211–32.

Valentine, G. (1996) '(Re)negotiating the heterosexual street' in N. Duncan (ed.) *Bodyspace* (London: Routledge) pp. 51–63.

Valentine, G. (2001) *Social Geographies* (Harlow: Prentice Hall).

Viles, H. (2005) 'A divided discipline?' in N. Castree, A. Rogers and D. Sherman (eds) *Questioning Geography* (Oxford: Blackwell) pp. 80–102.

von Engelhardt and J. Zimmerman (1988) *Theory of Earth Science* (Cambridge: Cambridge University Press).

Wade, P. (2002) *Race, Nature and Culture* (London: Pluto).

Wark, M. (1994) 'Third nature', *Cultural Studies*, 1: 115–32.

Watts, M. (1983) 'On the poverty of theory' in K. Hewitt (ed.) (1983) *Interpretations of Calamity* (Boston, Mass.: Allen & Unwin) pp. 231–62.

Weale, S. (2001) 'Grey area', the *Guardian* 23 January, 2–3.

Weeks, J. (1991) *Against Nature: Essays on History, Sexuality and Identity* (London: Rivers Oram).

Whatmore, S. (1997) 'Dissecting the autonomous self', *Society and Space* 15, 1: 37–53.

Whatmore, S. (1999) 'Culture-nature' in P. Cloke, P. Crang and M. Goodwin (eds) *Introducing Human Geographies* (London: Arnold) pp. 4–11.

Whatmore, S. (2003) *Hybrid Geographies* (London: Sage).

White, G. (1945) *Human Adjustment to Floods* (Chicago, Ill.: University of Chicago, Dept. of Geography Research Paper).

Willems-Braun, B. (1997) 'Buried epistemologies: the politics of nature in (post-) colonial British Columbia', *Annals of the Association of American Geographers* 87, 1: 3–31.

Williams, R. (1977) *Marxism and Literature* (Oxford: Oxford University Press).

Williams, R. (1980) *Problems of Materialism and Culture* (London: Verso).

Williams, R. (1983) *Keywords* 2nd edn (London: Flamingo).

Wilson, A. (1992) *The Culture of Nature* (Oxford: Blackwell).

Wilson, E.O. (ed.) (1988) *Biodiversity* (Washington, DC: Smithsonian Institute).

Winch, P. (1958) *The Idea of a Social Science and its Relation to Philosophy* (London: Routledge & Kegan Paul).

Wise, S. (2000) *Rattling the Cage* (New York: Profile Books).

Wittgenstein, L. (1922) *Tractatus Logico-Philosophicus* (London: Routledge & Kegan Paul).

Wolch, J. and J. Emel (eds) (1998) *Animal Geographies* (London: Verso).

Woods, R. (1986) 'Marx, Malthus and population crises' in R. J. Johnston and P. Taylor (eds) *A World in Crisis?* (Oxford: Blackwell) pp. 127–49.

Woods, M. (1998) 'Mad cows and hounded deer: political representations of animals in the British countryside', *Environment and Planning A* 30, 9: 1219–34.

Woolgar, S. (1988) *Science: The Very Idea* (Chichester: Ellis Horwood).

Women and Geography Study Group (1984) *Geography and Gender* (London: Hutchinson).

Wright, J.K. (1947) '*Terrae incognitae*: the place of the imagination in geography', *Annals of the Association of American Geographers* 37: 1–15.

Yatsu, E. (1988) *The Nature of Weathering* (Tokyo: Sozosha).

Young, I.M. (1990) *Throwing Like A Girl* (Bloomington, Ind.: Indiana University Press).

Young, O.R. (1994) *International Governance: Protecting the Environment in a Stateless Society* (Ithaca, NY: Cornell University Press).

Zimmerer, K. (1994) 'Human geography and the "new ecology" ', *Annals of the Association of American Geographers* 1: 108–25.

Zimmerer, K. (1996) 'Ecology as cornerstone and chimera in human geography' in C. Earle, K. Mathewson and M. Kenzer (eds) *Concepts in Human Geography* (Lanham, Md.: Rowman & Littlefield) pp. 161–88.

Zimmerer, K. (2000) 'The reworking of conservation geographies: non-equilibrium landscapes and nature-society hybrids', *Annals of the Association of American Geographers* 90, 2: 356–70.

Zimmerer, K. and K.R. Young (eds) (1998) *Nature's Geography: New Lessons for Conservation in Developing Countries* (Madison, Wisc.: University of Wisconsin Press).

Zimmerer, K. and T. Bassett (eds) (2003) *Political Ecology: An Integrative Approach to Geography and Environment-Development Studies* (New York: Guilford).

AUTHOR INDEX

Subject index

eBooks

eBooks – at www.eBookstore.tandf.co.uk

A library at your fingertips!

eBooks are electronic versions of printed books. You can store them on your PC/laptop or browse them online.

They have advantages for anyone needing rapid access to a wide variety of published, copyright information.

eBooks can help your research by enabling you to bookmark chapters, annotate text and use instant searches to find specific words or phrases. Several eBook files would fit on even a small laptop or PDA.

NEW: Save money by eSubscribing: cheap, online access to any eBook for as long as you need it.

Annual subscription packages

We now offer special low-cost bulk subscriptions to packages of eBooks in certain subject areas. These are available to libraries or to individuals.

For more information please contact webmaster.ebooks@tandf.co.uk

We're continually developing the eBook concept, so keep up to date by visiting the website.

www.eBookstore.tandf.co.uk